GENETICALLY MODIFIED PLANTS

GENETICALLY MODIFIED PLANTS: ASSESSING SAFETY AND MANAGING RISK

GEORGE T. TZOTZOS

GRAHAM P. HEAD

ROGER HULL

AMSTERDAM • BOSTON • HEIDELBERG • LONDON • NEW YORK • OXFORD
PARIS • SAN DIEGO • SAN FRANCISCO • SINGAPORE • SYDNEY • TOKYO

Academic Press is an imprint of Elsevier

Academic Press is an imprint of Elsevier
30 Corporate Drive, Suite 400, Burlington, MA 01803, USA
525 B Street, Suite 1900, San Diego, CA 92101-4495, USA
32 Jamestown Road, London, NW1 7BY, UK

First edition 2009

The views expressed herein are those of the authors and do not necessarily reflect
the views of their institutions, namely the United Nations Industrial Development
Organization, Monsanto Company and The John Innes Centre.

Library of Congress Cataloging-in-Publication Data
A catalog record for this book is available from the Library of Congress

British Library Cataloguing-in-Publication Data
A catalogue record for this book is available from the British Library

ISBN: 978-0-12-374106-6

For information on all Academic Press publications
visit our website at www.elsevierdirect.com

Typeset by Macmillan Publishing Solutions
www.macmillansolutions.com

Working together to grow
libraries in developing countries

www.elsevier.com | www.bookaid.org | www.sabre.org

ELSEVIER   BOOK AID
International   Sabre Foundation

Contents

Note: Colour insert appears after the index. The e-book for this title, including full-colour images, is available for purchase at www.elsevierdirect.com

Abbreviations

AIA	Advance Information Agreement
BC(1)	Back cross (1st generation)
BCH	Biosafety Clearing House
Bt	*Bacillus thuringiensis* (toxin)
CBD	Convention on Biodiversity
DNA	Deoxyribonucleic acid
DUS	Distinctiveness, Uniformity and genetic Stability
EFSA	European Food Safety Authority
EPA (US)	Environmental Protection Agency (of the USA)
ERA	Environmental Risk Assessment
EU	European Union
F(1)	Filial (1st generation)
FAO	Food and Agriculture Organization (of the United Nations)
GATT	General Agreements on Tariffs and Trade
GDP	Good developmental practice
GM	Genetically Modified (manipulated)
GMO	Genetically modified (manipulated) organism
GOI	Gene of interest
IAEA	International Atomic Energy Agency (of the United Nations)
ILSI	International Life Sciences Institute
LD_{50}	50% lethal dose
LMO	Living modified organism
NBC	National Biosafety Committee
NBF	National Biosafety Framework
NEPA (US)	National Environmental Policy Act (of the USA)
NGO	Non-governmental organization
NIH (US)	National Institutes of Health (of the USA)
NRC (US)	National Research Council (of the USA)
NTO	Non-target organism
OECD	Organization for Economic Cooperation and Development
OSTP (US)	Office of Science and Technology Policy (of the USA)
PNT	Plant with Novel Traits
RAC (US)	rDNA Advisory Committee (of the USA)

rDNA	recombinant DNA
RNA	Ribonucleic acid
SPS	Sanitary and phytosanitary measures (of the WTO)
TBT	Technical Barriers to Trade
UNEP-GEF	United Nations Environment Program – Global Environment Facility
VCU	Value for Cultivation and Use
WTO	World Trade Organization

Preface

For over ten thousand years, humans have tried to control their food supplies through agriculture, moving from being hunter-gathers to selecting and improving wild species of plants and animals. Over most of this time, humans have been successful in maintaining their food supplies, due initially to improving agricultural techniques and more recently to the application of scientific advances in subjects such as genetics and agronomy. With rapidly increasing populations and major changes in climatic conditions, food supply and security will be an increasingly important problem.

The introduction of genetically modified (GM) crops has been a major technological advance to world agriculture over the last few decades. It has been shown that this technology has the potential to help increase world food production at a time of rapid change.

Genetically modified organism (GMO) regulation plays a critical role, not only in the commercialization of GM crops, but also in research and development. This book is aimed at people involved in GMO regulation in government and industry, biosafety professionals dealing with GMO releases, and also those researching, producing and handling GM crops, e.g., agronomists, plant biologists, and molecular biologists, especially in developing countries. It should also be of use in undergraduate and graduate courses on GM technology and to the general public who wish to be informed on the subject.

The book reviews current and future applications of GM technology, provides a background on GMO regulation, and critically analyses differences among different regulatory systems. The first chapter sets the scene by describing features of conventional crop breeding and production, the constraints on food production, the basics of GM technology, and the current and future applications of the technology. Because the aim of the book is risk assessment and management, the second chapter deals with the basics of risk assessment and risk management. The main body of the book is dedicated to reviewing assessment procedures concerning potential risks to human health (Chapter 3) and the environment from releases of GM plants (Chapter 4), with illustrative examples. The growth of GM technology has concerned the public and decision-making on GM releases not only involves the scientific assessment of potential risks, but also public perception of these factors – this is discussed in Chapter 5. Chapter 6 describes the national and international regulatory structures that have been developed to handle this technology. Two of the Appendices (B and C) outline in detail the information that is required by many regulatory authorities for assessing proposed GM releases, and should provide guidance for new authorities who are developing their regulatory systems. Appendix D gives five case studies on releases which highlight specific issues.

Other features include:

- Baselines against which risk assessment is made;
- Grey areas in risk assessment procedures;
- Inputs into decision-making on GMO releases;
- How public perception of biotechnology impinges on GM regulation;
- The major international issues pertaining to GMO releases.

This book would not be possible without the generous input from numerous people who have provided advice and facilities. GT wishes specifically to thank his wife, Susan; GH thanks Natalia for her support and understanding; RH specifically thanks the John Innes Centre for access to their library and computing facilities, and his wife, Jenny, who has tolerated "piles of paper" on any clear surface around the house.

George Tzotzos, Graham Head and Roger Hull
May 2009

1

Setting the Context: Agriculture and Genetic Modification

ABSTRACT

This chapter sets the scene for the development of genetic modification of crops and associated biosafety issues to provide a background for the risk assessment, mitigation and management topics which are described in subsequent chapters. We discuss why we need genetic modification of crops, the genetic modification traits that have been released to the field thus far, and the traits which are likely to be released in the near future. We also discuss the type of biosafety concerns raised by genetic modification technologies and the baselines against which risk assessments are made.

OUTLINE

I. AGRICULTURE

A. History

Agriculture started more than 10,000 years ago (Fig. 1.1) when humans moved from being hunter/gatherers towards organized societies with food producers and food consumers. Initially, wild plant species were

Years before present (BP: 2009)

10,000	Domestication of cereal grains, major legumes and root crops, vegetable, oil, fibre and fruit crops
	Spread of crops within Old World Spread of crops within New World Spread of crops within Asia
2,000	Domestication of forage crops and drug sources
	Spread of crops within Africa Spread of crops from Asia to Europe
500	Spread of crops from New World to Old World Worldwide dispersal of crops
200	Domestication and dispersal of further crops, e.g. rubber, sugar beet, blueberries, macademia nuts
100	Understanding of genetics leading to directed plant breeding
c 60	Development of new techniques (e.g., mutagenesis) in plant breeding
50	Understanding of DNA
20	Genetic modification of plants
15	Release of genetically modified plants to the field

FIGURE 1.1   Timeline of agriculture.

domesticated by choosing those variants that had good properties (e.g., yield, reliability, lack of toxins and ease of cultivation). This was followed by a long period of time in which farmers undertook empirical crop breeding by selecting the best variants, cross-fertilizing them and then further selecting to improve them. Obviously, this empirical crop breeding could only be done to locally available varieties and species and thus there were centres of domestication/breeding associated with centres of origin and diversity of the various crop species (Box 1.1).

BOX 1.1

CENTRES OF ORIGIN OF CROPS

The early domestication of crops occurred in their centres of origin. Vavilov (1935) identified eight primary centres of origin of the world's major crops (Fig. 1).

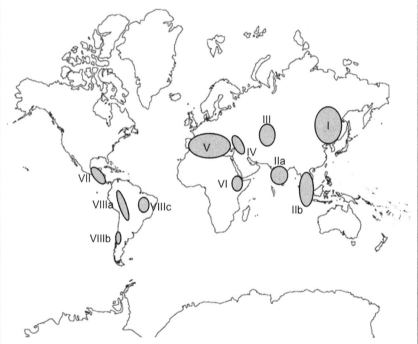

FIGURE 1   Vavilov's primary centres of origin of the world's major crops (see colour section).

I   **The Chinese Centre:** Crops from this area include: buckwheat, soybean, Chinese yam, radish, Chinese cabbage, onion, cucumber, pear, peach, apricot, cherry, walnut, sugarcane and hemp.   *(cont'd)*

BOX 1.1 *(cont'd)*

IIa **The Indian Centre** is the centre for crop species including: rice, chickpea, cowpea, eggplant, cucumber, radish, taro, yam, mango, orange, sugarcane, coconut palm, sesame, black pepper and bamboo.

IIb **The Indo-Malayan Centre** is the centre for crop species including: velvet bean, banana, mangosteen, coconut palm, sugarcane and black pepper.

III **The Inner Asiatic Centre** is the centre for crop species including: wheat, pea, lentil, horse bean, chickpea, mustard, flax, cotton, onion, garlic, carrot, pear, almond, grape and apple.

IV **The Asia Minor or Near-Eastern Centre** is the centre for crop species including: wheat (einkorn, durum and common), rye, oats, lentil, lupine, alfalfa, fig, apple, pear and cherry.

V **The Mediterranean Centre** includes crop species such as: durum wheat, emmer, spelt, pea, lupine, various clovers, flax, oilseed rape, olive, beet, cabbage, turnip, lettuce, parsnip and hop.

VI **The Abyssinian (now Ethiopian) Centre** contains crop species such as: emmer, barley, pearl millet, cowpea, flax and coffee.

VII **The South Mexican and Central American Centre** contains crop species such as: maize, common bean, lima bean, winter pumpkin, sweet potato, papaya, cherry tomato and cacao.

VIII The South Andes region comprises three centres:

VIIIa **Peruvian, Equadorean, Bolivian Centre** contains crop species including: various potato species, lima bean, common bean, tomato, pumpkin, Egyptian cotton and tobacco.

VIIIb **The Chilean Centre** contains crop species including: common potato and wild strawberry.

VIIIb **The Brazilian–Paraguayan Centre** is important for: manioc (cassava), peanut, rubber tree, pineapple and Brazil nut.

While the primary domestication of the various crops occurred at these centres of origin there are various other centres of diversification. For instance, banana has been diversified in the Indo-Malayan region, in East Africa and in West Africa.

The centres of origin and diversification are important sources for genes for plant breeding as they contain the wild species and landraces used by indigenous farmers. There can be specific biosafety considerations for the release of GM crops in these regions.

The domestication and crop breeding led to populations of crop species known as landraces (see the list of commonly used terms in Appendix A) adapted to specific regions and environments.

Up to the 16th century these domesticated crops were dispersed within continents by the migration of people and by local trade. With the discovery of the New World and increasing intercontinental trade, crops were dispersed worldwide to other climatic areas where they could be grown.

Over the last 200 years or so, increases in human population and urbanization have led to agriculture dividing even more markedly between food producers and food users. These and other factors have resulted in a requirement for increased and consistent food supplies with concomitant changes in agricultural practices and the stronger recognition of the constraints on crop production.

Crop agriculture covers about 11% of the world land area. Essentially, crop agriculture involves the capture of solar energy to give a range of products that are used for human food, animal feed and, very recently, industrial processes such as the production of ethanol. Animals also are fed by grazing which uses a further 22% of the world land area; thus food and feed involves about one-third of the global land area. The proportion of crop production used for food versus feed varies among different societies, with the more affluent having diets containing more animal products. Another source of animal feed is plant crop products (particularly from corn and soybeans) and consequently more affluent societies use more crop products per capita because the conversion of plant products to animal products is relatively inefficient. Thus, requirements for crop production increase not only with population size but also with the affluence of societies. As well as crop production, humans also cultivate trees (silviculture or forestry) and fish (aquaculture), both of which are seeing applications of modern biotechnology.

Overall, the world produces enough food to feed the current population of about 6.3 billion. However, it is estimated that the world population will increase to about 9 billion by the year 2050 and will need about twice the current food production. Most of the population increase will be in urban areas with larger cities taking away potential arable land, and using more water. Furthermore, climate change is already affecting the potential for increasing crop yields in several major crop production areas and this problem is predicted to get worse. Thus, there will be an increasing need to produce more food from less arable land area.

There are three major constraints on crop production: biotic, abiotic and genetic (Box 1.2). *Biotic* constraints are losses due to pests, diseases and competing plants (weeds), and are estimated worldwide to reduce yields by about 40%. *Abiotic* losses include climatic factors such as drought and temperature and soil factors such as fertility and salinity. *Genetic* factors include the potential in the germplasm of that crop to increase yield and the quality of the product.

BOX 1.2

CONSTRAINTS ON CROP PRODUCTION

Biotic constraints include:

- *Insect pests* such as stem borers, bollworms, aphids and whiteflies. These can cause losses due to the feeding of the larval stages and as vectors of virus diseases.
- *Diseases* caused by fungi, bacteria and viruses. Sometimes these kill the crop plant (e.g., potato blight) but often they debilitate the crop.
- Weeds which cause losses by competition with the crop for food and water; some weeds (e.g., Striga) parasitize the crop plant, taking nutrients from it and killing it.

Abiotic constraints include:

- Drought, flooding, high or low temperatures, soil fertility, salinity, toxic metals.

Genetic constraints include:

- Non-availability of desired traits (e.g., disease resistance, yield, plant architecture) in sexually compatible species for conventional breeding.

II. DEALING WITH CONSTRAINTS

Currently, the main ways of dealing with crop production constraints are to improve agronomic practices and to breed in characters which mitigate the constraints.

A. Agronomic Practices

In many farming systems, biotic constraints are controlled by the application of pesticides, e.g. insecticides, fungicides and herbicides; the application of insecticides and fungicides is not limited to the growing crop but is also used to prevent losses in crop storage. Although these often prove effective in the short term, several major problems can arise. Because of natural variation in target pest and weed populations, resistance to the pesticide can arise which means that a new pesticide has to be deployed. There are also adverse impacts on human health, both through poisoning of farmers because of insufficient protection and pesticide residues in the product which can affect the consumer and necessitate testing systems and quality control. There are estimated to be between 2 and 5 million pesticide poisonings per year, of which 40,000 are fatal (http://www.fao-ilo.org/fao-ilo-safety/en/).

Similarly, there are potential adverse impacts on the environment through non-target effects of the pesticide either killing off non-target species, or affecting systems such as food chains of birds. In some farming systems, attempts are made to control biotic constraints by integrated pest management strategies which endeavour to reach a natural balance of prey and predator. The success of this approach is subject to the vagaries of climatic and other factors and has limited application to the large monocultures which are necessary for feeding large urban populations.

Abiotic constraints can also be controlled by agronomic practices. For instance, in some farming systems the effect of drought is controlled by irrigation. With the current burgeoning of the world population this is leading to problems of how to allocate water supplies to humans, their domestic animals and crop production. Furthermore, overuse of irrigation can lead to salinization of the soil. As well as using herbicides, weeds are frequently controlled by ploughing the soil which is causing soil erosion in many countries. Another example of agronomic practices addressing potential abiotic constraints is the use of short varieties of cereals to minimize storm damage causing "lodging" of the crop.

Constraints to yield are also addressed by the application of fertilizers. This has led to the breeding of crop varieties that utilize fertilizers, especially nitrogen, more efficiently.

Climate change is likely to lead to an exacerbation of both biotic and abiotic constraints worldwide and especially in tropical developing countries. There are strong links between some modern agricultural practices and factors that may lead to climate change. For instance, the production and application of inorganic fertilizers involve the use of fossil fuels and energy thus increasing the environmental footprint of agriculture.

B. Breeding

The rediscovery of the work of Gregor Mendel in the early 1900s enabled plant breeding to be put on a scientific basis (for a detailed description of Mendel's work and the subsequent development of classical breeding, see Simon Mawer (2006)). The basis of classical breeding programmes for crops is variation – useful characters or traits that can be bred into the crop to overcome specific constraints such as pests and diseases or to improve the agronomic performance of that crop in terms of yield or food quality. Variation is the result of mutation or genetic recombination. In recombination, pieces of DNA exchange between chromosomes, usually during meiosis (see Appendix A).

Breeding relies on the exchange of genetic material between the parents and usually involves the parents being sexually compatible; sexual compatibility usually occurs within species and not between them. Although there is natural genetic variation within species (Fig. 1.2), often

FIGURE 1.2   Variation in *Brassica oleracea*. In the centre is the wild-type *B. oleraces* surrounded by cultivated variants with enlarged terminal bud (cabbage), coloured variety of swollen terminal bud, swollen axillary buds (Brussels sprouts), swollen flower heads (cauliflower and broccoli) and swollen stem (kohl rabi) (see colour section).

there is not the required trait, such as disease resistance. Thus, various ways of overcoming this genetic constraint of lack of suitable characters in sexually compatible species have been developed, increasing the available variation and thus the scope for genetic exchanges between species.

The variation within species can be increased by artificial mutation (see Appendix A). Mutagenesis, either by chemicals or by radiation has been widely used since the 1950s (Box 1.3). It has been estimated that about 80% of the most common varieties of the major crop species have artificial mutagenesis in their parentage. For a list of primary artificial mutants, see the FAO/IAEA Mutant Variety DataBase (www-mvd.iaea.org).

To increase the availability of genetic variation, "wide crosses" between species which are not normally sexually compatible can be undertaken using techniques such as embryo rescue or protoplast fusion (Box 1.4). Embryo rescue has been used to produce a new fertile rice variety (NERICA) from an intraspecific cross between *Oryza sativa* and *O. glaberrima*; NERICA has disease resistance and some good agronomic properties lacking in *O. sativa*. Interspecific crosses can occur naturally

---

BOX 1.3

ARTIFICIAL MUTAGENESIS

Mutagenesis of plants can be caused artificially by either chemical or physical treatment.

Chemical mutagenesis is usually undertaken by treatment with either ethyl methanesulphonate (EMS) or diethyl sulphate (DES). Such treatment causes chemical modification of nucleotides in a random manner. EMS usually mutates cytidine residues to thymidine (C > T mutations). Problems with chemical mutagenesis include uncertain penetration of the target plant material and high rates of mutation when penetration occurs.

Physical mutagenesis (also termed nuclear plant breeding) is by X-rays, gamma rays or fast neutrons which modify nucleotides in a random manner. This treatment gives good penetration and good dose control but can cause chromosomal aberrations.

Targets for mutagenesis for seed propagated crops are usually dry seed (which can have problems with the formation of chimaeras) or gametes. For vegetatively propagated crops, tubers, dormant buds or shoot tips are used.

Analysis of the data on rice in the FAO/IAEA database (www-mdv.iaea. org/MDV) showed that 89% of the 498 mutated varieties were produced by radiation mutagenesis, 2% by chemical mutagenesis and 9% by other means; these are primary mutants and many will have been used in conventional breeding programmes. The traits produced by mutation included plant characteristics such as plant height, leaf morphology and tillering, agronomic characteristics such as earliness, yield, fertilizer response, cold tolerance and disease resistance and seed characteristics such as grain morphology.

---

but rarely as shown by the parentage of modern wheat (Fig. 1.3). These crosses were selected by farmers.

As with agronomic practices, the genetic control of biotic constraints such as insect or disease resistance can be overcome by natural variants of the pest. Thus, plant breeders are constantly seeking new sources of pest resistance.

The flow chart for a conventional plant breeding programme is shown in Fig. 1.4.

Breeding programmes for elite varieties can be very complex as shown by that for the rice variety IR64 (Fig. 1.5). The breeding for this variety included input of traits from various landraces, sexually compatible *Oryza* species and other varieties including ones that had been produced using mutagenesis.

At each stage in such a breeding programme, the plant with the desired trait is usually crossed with another variety (or species) of the

---

BOX 1.4

WIDE CROSSES

Normally, crosses between different species (interspecific) or different genera (intergeneric) do not succeed because of sexual incompatibility. Usually, this incompatibility can show at one of two stages: in the female parent of the cross where there is abortion due to parental mutual incompatibility or in the progeny which are infertile. There are various techniques which breeders use in attempts to overcome these problems including:

- **Embryo rescue**: Prior to abortion, embryos are excised and grown *in vitro* on a suitable medium.
- **Chromosome doubling**: Sterility occurs because the first generation of a wide cross contains an uneven number of unmatched chromosomes. To overcome this problem, the number of chromosomes is doubled using the cell division inhibitor, colchicine.
- **Protoplast fusion**: Protoplasts, which are cell wall-free plant cells (that is plant cells lacking the cell wall) from different species or genera are fused together (often by treatment with an electric current) and then are grown *in vitro* to form hybrid plants. These usually need chromosome doubling to become fertile.

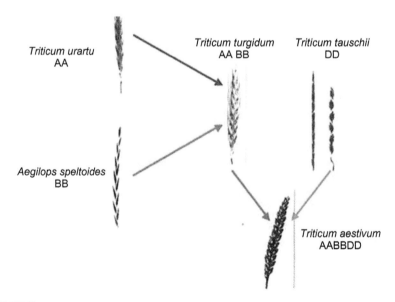

FIGURE 1.3   Evolution of modern wheat (triticum aestivum) arising from natural interspecific crosses of wild grass species bringing together three genomes (see colour section).

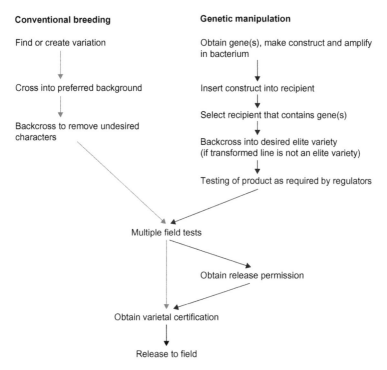

Conventional breeding

Find or create variation

Cross into preferred background

Backcross to remove undesired characters

Genetic manipulation

Obtain gene(s), make construct and amplify in bacterium

Insert construct into recipient

Select recipient that contains gene(s)

Backcross into desired elite variety (if transformed line is not an elite variety)

Testing of product as required by regulators

Multiple field tests

Obtain release permission

Obtain varietal certification

Release to field

FIGURE 1.4 Flow charts for production of a new variety by conventional breeding and genetic manipulation.

crop to produce an F1 (filial) generation comprising a large number (up to 10s of thousands) of heterozygous (see Appendix A) progeny. Progeny that contain the desired trait are selected but they also may contain undesirable characters from the non-elite variety. Selected progeny are then backcrossed (see Appendix A) to the elite variety to give a BC1 generation, and selected again and further backcrossed until a new stable homozygous (see Appendix A) crop line containing the characters from the elite line parent as well as the new desired trait is obtained. This may take up to 10 backcrosses but the time involved can be shortened by techniques such as marker-assisted breeding (see Appendix A) and, for annual species, by growing two crops a year, one in a northern country and one in a southern country. Even so, the process can take a long time and involves growing large numbers of plants. The new line is then field tested under the range of conditions where it is expected to be grown to determine if the new combination of genes is stable and durable. In many countries, most new varieties of a crop produced by conventional breeding must be accepted for the National List before the farmer can use them. For instance, to be added to the National List in a European country, a new variety is tested for its Distinctiveness (i.e. differing from

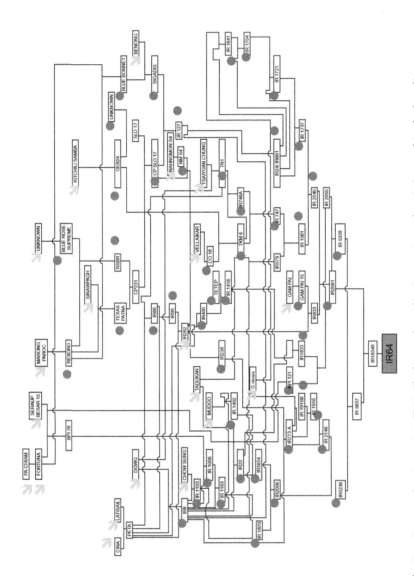

FIGURE 1.5 Breeding tree for rice variety IR64. The boxes show the rice species and varieties used in the breeding programmes. The yellow arrows indicate landraces and the red dots indicate new varieties derived either from breeding programmes or by mutagenesis (see colour section).

BOX 1.5

F1 HYBRID VARIETIES

An F1 (filial 1) hybrid variety or cultivar is usually from two different parent cultivars, each of which are inbred for a number of generations to the extent that they are almost homozygous. The divergence between the parent lines promotes improved growth and yield characteristics through the phenomenon of "hybrid vigour" (termed heterosis), while the homozygosity of the parent lines ensures a phenotypically uniform F1 generation. Thus, F1 hybrid seed produces a homogeneous, predictable and higher yielding crop than either of its parents. However, in the progeny from an F1 hybrid (the F2 generation) the parental genes segregate producing a mixture of plants, thus losing the uniformity and hybrid vigour of the F1 hybrid. Thus, F1 hybrid seed has to be produced for each cropping season and the farmer has to buy new seed and not save it from his previous crop.

---

existing varieties of that crop), its *U*niformity and its genetic *S*tability (the DUS tests) and, for agricultural crops, it must have satisfactory *V*alue for Cultivation and *U*se (VCU). The DUS tests and VCU trials are carried out at approved centres covering the expected range of agronomic conditions expected for that crop usually over a two year period and using protocols and procedures approved by the EU Plant Variety and Seeds Committee. Similar forms of varietal registration occur in most countries.

One breeding technique that increases yield in certain crops is the production of F1 hybrids (Box 1.5).

III. GENETIC MANIPULATION (GM) TECHNOLOGY

Over the last three decades advances in the ability to manipulate DNA have led to ways of bypassing the main limitation of plant breeding – that of lack of suitable genes in sexually compatible species. This technology is called genetic manipulation or genetic modification (GM) and has been defined:

Genetic Manipulation is a technology in which a gene [see Appendix A] or genes are taken from one organism (the donor) or are synthesised de novo, possibly modified and then inserted into another organism (the recipient) in an attempt to transfer a desired trait or character.

The technology is also called genetic engineering, recombinant DNA technology and bioengineering among other terms. It is a new technology toolkit among the many that comprise biotechnology such as brewing, production of enzymes by fermentation and marker-assisted breeding.

A. The Basic Technology

A flow chart for the production of a GM crop plant is shown in Fig. 1.4. The desired gene or genes are obtained from a *donor* and are inserted into the *recipient* through a process called transformation (see Appendix A); these processes are described in more detail in Sections III.B and III.C. The transformant or transformants that contain the desired gene(s) are selected and grown into full plants and are backcrossed to ensure homozygosity as described above for conventional breeding. This produces several lines which are tested to determine if the trait (see Appendix A) or traits that were introduced have the desired properties. The desirable lines are then subjected to a wide range of tests to obtain information required by regulators (see Chapters 3 and 4) before being bulked up, certified as a new variety and released to the field.

Each instance of a GM organism that is proposed for commercial release is termed an event (see Appendix A). For example, the same gene inserted into a given plant genome at two different locations (i.e., loci) in that plant's DNA would be considered two different "events". Similarly, two different genes inserted at the same locus of two same-species plants would also be considered two different "events".

B. Gene Identification and Manipulation

Current GM plants contain one or more conventional gene(s) (see Appendix A) (we will use the term "conventional genes" to describe those that express as protein to contrast them from "silencing genes" that do not express proteins (see below)). However, recent discoveries have shown that plants have a defence system that degrades "foreign" nucleic acids (termed gene silencing (see Appendix A) or RNA interference, RNAi) (Box 1.6). Thus, new traits can also be produced by the insertion of non-coding sequences. Genes for transformation can be obtained from a wide range of sources (Table 1.1).

If a gene has been obtained from another variety of the same species, the transformant is termed cisgenic (see Appendix A), as compared to a transgenic transformation in which the gene is taken from a different species. The sequence of a gene from a very different organism may be modified to ensure that it functions well in plant cells.

Genes are usually identified by their DNA nucleotide sequence which is checked to ensure that it does not contain any unwanted aberrations.

The gene of interest (GOI) is assembled into a construct which comprises all the elements required for successful transformation (Fig. 1.6).

The construct is usually a plasmid (see Appendix A) so that it can be amplified in bacteria to provide enough material for transformation. The GOI requires a promoter (see Appendix A) and a termination sequence

BOX 1.6

RNA SILENCING

RNA silencing, also known as gene silencing, transcriptional gene silencing, post-transcriptional gene silencing and RNA interference (RNAi), is a plant defence mechanism against "foreign" nucleic acids. It is also found in insects, fungi (where it is known as gene quelling), nematodes and mammals.

A basic feature of RNA silencing in all these organisms is that it involves highly structured double-stranded RNA. This unusual molecule is recognized as being "foreign" and is processed through a pathway that is shown in Fig. 1.

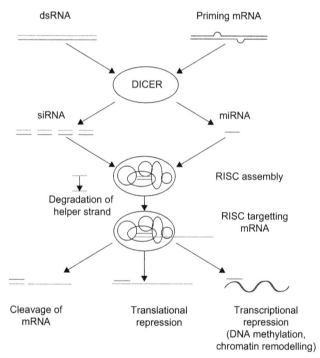

FIGURE 1 RNA silencing pathways. The red line indicates the guide strand and the green line the anti-guide (helper) strand and the target mRNA. The left-hand pathway is that for mRNAs and the right-hand pathway for micro-RNAs which are involved in plant development. Both pathways pass through Dicer which cleaves the double-stranded RNA to small fragments and through RISC which amplifies the system. From Hull (2009) with kind permission of Elsevier (see colour section).

The double-stranded RNA is cleaved by an enzyme (Dicer) to give small fragments (small interfering RNA, siRNA) and the strands of the siRNA are separated into the guide strand which is complementary to the cognate messenger

(cont'd)

BOX 1.6 (*cont'd*)

RNA (mRNA) and the anti-guide RNA which is degraded. The guide RNA is then incorporated into the RNA-induced silencing complex (RISC) which targets it to the cognate mRNA, forming a duplex with that RNA. This then is cleaved by RISC. Thus, the process is reiterative and the presence of siRNA molecules primes the defence system against further cognate mRNAs.

This defence system is currently used in two ways in genetic modification of plants. Most plant viruses have genomes of single-stranded RNA (mRNA) and constructs can be introduced into plants to prime the defence system against the virus. Similarly, constructs can be directed against endogenous plant mRNAs to turn off specific genes (e.g., those that cause food allergies). Such constructs are designed as hairpins to be transcribed to give double-stranded RNA (Fig. 2) which would prime the defence system.

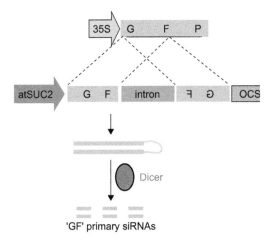

FIGURE 2   Construct for RNA silencing gene. This example shows a construct for silencing the expression of green fluorescence protein (GFP). The top line shows the GFP gene used as a marker in a plant. The second line shows the silencing construct with a promoter (red arrow on left), portions of the GFP sequence in positive and reverse sense separated by an intron and on the right the terminator. The next two lines show how the RNA expressed from the construct forms a double-stranded molecule which is processed to form siRNAs by Dicer (see colour section).

The risk assessment of transformants containing RNA silencing constructs is different to those containing conventional genes in that no protein is expressed. Thus, properties of a transgene protein, such as allergenicity and toxicity, are not relevant. However, the trait conferred by the silencing construct and unintended effects should be considered in the same way as conventional gene constructs.

TABLE 1.1 Sources of genes.

- Another variety of the same species
- Another species of the same genus
- Another genus within the same family
- Another family within the same kingdom (i.e., plants)
- Another kingdom (animal, fungus, bacteria, etc.)
- Synthesized *de novo*

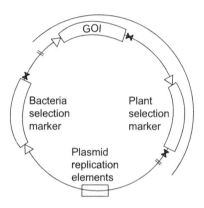

▷ Promoter

✗ Termination sequence

= T-DNA left and right borders

FIGURE 1.6 Typical plasmid construct for transformation into plants. The boxes show genes (GOI = gene of interest) and sequences necessary for their expression. The open arrows indicate promoters, the double infilled arrows, termination sequences. For transformation using Agrobacterium, the T-DNA left and right borders are included; for biolistic transformation they are not needed.

(see Appendix A) to tell the recipient plant cell machinery where and when to start and finish expression (see Appendix A) of the gene. A promoter may be constitutive (expressing the gene in most, if not all, cells), tissue dependent (expressing only in certain tissues such as roots or seed) or inducible (expressing only when a certain inducer, usually a chemical, is applied).

As only a proportion of bacterial cells (in which the construct is amplified) and plant cells are transformed, additional genes are usually required for selection of successful transformants – these are termed selectable marker genes (Table 1.2). Because of differences in the way that bacterial and plant cells function, different selection marker genes may be required for each organism. For replication (and thus amplification) in the bacterial host the construct requires specific DNA sequences. If the transformation of the plant cells is going to be by *Agrobacterium* (see Section III.C.2), the GOI and plant selection markers have to be delimited by the left and right border sequences.

C. Plant Transformation Techniques

There are several techniques for introducing genes into plant chromosomes (Table 1.3) but only two – biolistics and *Agrobacterium* – are widely used.

TABLE 1.2   Examples of selectable marker genes.

B indicates that selection operates in bacteria, P selection operates in plants

Antibiotic resistance

nptII gene (resistance to kanamycin and neomycin) (B, P)
aphIV gene (resistance to hygromycin) (B, P)
bla gene (resistance to ampicillin) (B, P)
aad gene (resistance to streptomycin and spectinomycin) (B, P)

Herbicide tolerance

bar gene (tolerance to phosphotricine, glyphosinate, bialophos) (P)
ALS gene (tolerance to chlorosulpharan) (P)

Metabolic genes

Phosphomannose isomerise (PMI) gene metabolizes mannose to give fructose (P)

Other markers

Green fluorescent protein (GFP) (P)
B-galactosidase (GUS) (P)

TABLE 1.3   Transformation techniques.

Direct transfer of gene to plant chromosome

Biolistics
Electroporation
Direct uptake in protoplasts
Microinjection

Indirect transfer of gene to plant chromosome

Agrobacterium-mediated transformation

General transfer of many genes

Protoplast fusion

The basic transformation technique is shown in Fig. 1.7 and, because the technique can have biosafety implication (see Chapters 3 and 4), the two most widely used ones are described here in more detail.

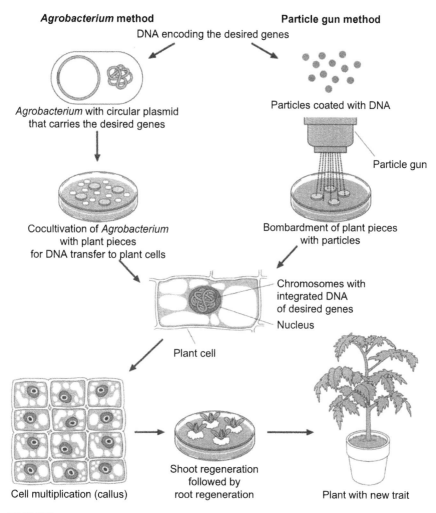

FIGURE 1.7 Schematic representation of two different methods to create transgenic plants. In the *Agrobacterium* method (left), DNA carrying desired genes is inserted into the Ti plasmid of the bacterium and, when the bacterium infects wounded tissue, this DNA is transferred to a cell nucleus and integrated into the chromosome. In the particle gun method, metal particles coated with DNA are fired into plant cells and the DNA becomes integrated into the plant chromosome. When a new plant is regenerated from a single plant cell, all the cells in the plant carry the new genes. From Chrispeels and Sadava (1994) with kind permission of the publishers Jones and Bartlett.

1. *Biolistics*

The biolistic procedure, also known as the "gene gun", involves firing small metal particles coated with the construct DNA into plant cells. The metal particles, usually gold or tungsten, are accelerated to high speed by the rapid release of high pressure helium in the gene gun into the target cells. The target is totipotent (see Appendix A) cells, usually consisting of an embryogenic suspension culture or an embryogenic callus derived from the recipient plant. Some of the metal particles enter the nuclei of cells without causing lethal damage and deliver the construct so that it can integrate by recombination with the chromosomes. The target cells are then induced to form plants under selection which only permits material expressing the selectable marker (and usually the GOI) to survive.

2. Agrobacterium-*Mediated Transformation*

Agrobacterium tumefaciens is a plant pathogenic bacterium that contains a plasmid (the tumour-inducing or Ti plasmid), part of which (the T-DNA) integrates into the host plant chromosomes (Box 1.7); it has been termed "nature's genetic engineer". For plant transformation, the tumour-inducing genes are removed to make a "disarmed" plasmid. The GOI is placed between the left and right borders of the T-DNA which is then reintroduced into *Agrobacterium*. This *Agrobacterium* is then co-cultivated with suitable target cells (embryogenic suspensions or callus) to enable the modified T-DNA to integrate into the cells. After removal of the *Agrobacterium*, the transformed cells are grown into plants under selection.

3. *Biosafety Issues Relating to Transformation Technique*

There are three basic biosafety issues relating to the transformation technique: variation in the number of inserts, variation in the copy number in inserts and potential for contamination. The biolistic techniques frequently lead to a relatively large number of independent inserts of the transgene into the plant chromosomes. Some of the inserts may contain deletions or insertions, and others may be in the reverse orientation to the GOI. On the other hand, *Agrobacterium* transformation usually results in one or a few insertions into the host chromosomes but these inserts may have multiple copies of the GOI. It is important to have information on the insert number and the copy number in making a risk assessment so that potential problems with multiple inserts and large copy numbers can be identified.

The biolistic technique will insert any DNA that is in the preparation used to coat the metal particles. Thus, any contaminant of the plasmid preparation amplified in bacteria could be inserted. It was thought that *Agrobacterium* only inserted sequences flanked by the T-DNA borders; however, it has recently been reported that fragments from the Ti plasmid can also be integrated (Ülker *et al.*, 2008).

BOX 1.7

AGROBACTERIUM TUMEFACIENS

Agrobacterium is a plant pathogenic bacterium that causes tumours (crown gall disease) in some plant species (Fig. 1A). The bacterium contains a plasmid (the tumour-inducing or Ti plasmid), part of which (the T-DNA) integrates into the host plant chromosomes (Fig. 1B). The Ti plasmid contains several genes including the *vir* genes which control the process of infection of the plant and transfer of the T-DNA to the chromosome. The T-DNA contains the tumour-inducing (auxin) genes and a gene that expresses specific compounds, opines, which are used by the bacterium as a carbon source. Thus the bacterium creates its own food supply within the plant.

A

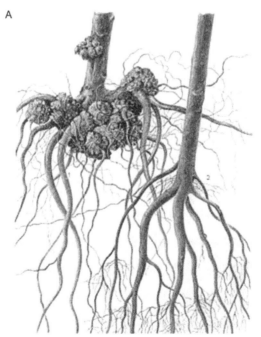

FIGURE 1   *Agrobacterium tumefaciens*. **A**. Infected plant showing crown gall. **B**. (see page 22.) The Ti plasmid showing coding and other regions. At the top the T-DNA region, bounded by the left and right borders, is indicated. In the unmodified plasmid this region contains genes for the maintenance of *A. tumefaciens* in its host; these genes are removed to make the plant transformation vector. The lower part of the genome contains genes and regions involved in the replication of the plasmid and in the process of plant infection by the bacterium (see colour section). Panel B from http://upload.wikimedia.org/wikipedia/commons/8/89/Ti_Plasmid.jpg.

(cont'd)

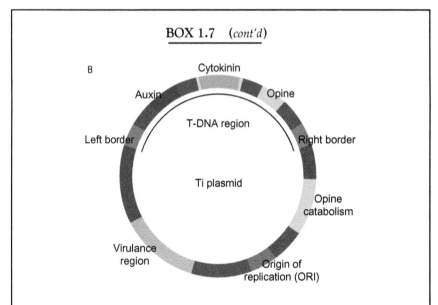

BOX 1.7   *(cont'd)*

The T-DNA is defined by specific sequences, the left and right borders, which are involved with the integration of the DNA into the host chromosome. The integration is at random sites and so can disrupt the normal functioning of the plant cell by integration into important genes or other sequences.

IV. CURRENT AND FUTURE GM CROPS

A. Current Crops

In 2008, 125 million hectares (309 million acres) were planted to GM crops worldwide (James, 2008) (for up-to-date information from ISAAA go to http://www.isaaa.org). There are currently four major GM crops (soybean, corn (maize), cotton and oil-seed rape (canola)) containing two major traits (insect resistance and herbicide tolerance) released to the field for commercial production. Initially, individual crop plus trait combinations were released but there is an increasing trend for crops to have two or more "stacked" traits in a single plant. By 2008, GM crops had been released in 25 countries (Fig. 1.8).

As well as these major crops, there have been minor commercial releases such as virus-resistant papaya and squash and carnations with modified flower colour.

As shown in Fig. 1.8, there has been a steady increase in the adoption of GM crops since 1996 with an increasing proportion being in developing countries. The adoption of GM crops is leading to changes in agronomic

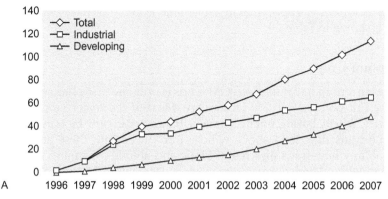

A

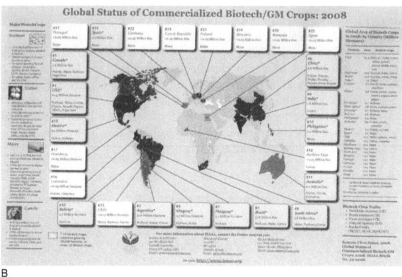

B

FIGURE 1.8 **A.** Global area of biotech crops 1996–2008: industrial and developing countries (million hectares). **B.** Countries growing biotech crops in 2007, 13 of which are termed "mega-countries", growing 50,000 hectares or more. From James (2007; 2008). With kind permission of the International Service for the Acquisition of Agrobiotech Applications (see colour section).

practices with agricultural practices being adapted to the GM traits. Among examples of this adaptation is no- or low tillage soil preparation for seed sowing linked with use of herbicide-tolerant crop varieties which helps to reduce soil erosion (see Box 4.9). Similarly, the use of insect-resistant crops has led to a significant reduction in the application of pesticides with a concomitant reduction in pesticide poisoning of farmers (Pray *et al.*, 2002). Additional benefits of GM crops include increased crop productivity (fewer losses to insects and weeds), more reliable crop production and increased farmer income. The impact that first generation GM crops have had over 11 years is shown in Table 1.4.

B. Future Crops

The current range of released GM crops is only the first phase of applying the full potential of this technology. Most of the major crop species (e.g., rice, wheat, barley, potato, cassava, banana) can now be transformed and the traits described above will be introduced into these and other crops. Many new traits directed at biotic and abiotic constraints are now in the pipeline. There are also five major areas where GM technology can play a major role: crop yield, food quality, post-harvest protection and processing of food, pharmaceutical products and industrial products. As well as increasing the amount and stability of food production, many of these traits will help to conserve biodiversity, increase the agricultural sustainability and the agro-industry potential of many countries and help mitigate some of the effects of climate change. GM has the potential to alter crop products in many ways and some examples are given below.

Some of the potential future GM products (e.g., trees, production of pharmaceuticals, transgenic microbes) will raise new biosafety concerns. As more knowledge and experience accumulates, it is likely that regulatory structures will evolve to address these.

1. Crop Yield

The GM approach is one important tool in attempts to fully realize the potential yield of current crop varieties. Transgenes are being developed to mitigate the effects of biotic constraints other than insects and weeds. For instance, strategies to protect potatoes against late blight fungus infections and nematode damage are being field tested. Protection against virus infection by using non-coding viral sequences to activate the natural plant defence by RNA silencing (Box 1.6) is being tested.

TABLE 1.4 Impact of GM crops 1996–2006. Data from PG Economics Ltd (2008).

	Yields		Change in inputs	
	Additional production (million tons)	Average trait impact – yield	Herbicide	Insecticide
Canola	0.2	+3.0%	−24.2%	−
Corn	9.6	+6.7%	−4.6%	−5.3%
Cotton	1.4	+15.0%	−14.5%	−24.6%
Soybean	11.6	+20.0%	−20.4%	−

There also is much work being directed against abiotic constraints with drought tolerant traits being available by about 2012. Linked with these traits are tolerance against heat, cold and salinity. A major aim is to reduce the requirement of crop plants for nitrogen fertilizers by incorporating gene systems that make the plant take up, and use, fertilizer more efficiently.

The GM approach also raises the possibility that the potential yield of a crop can be increased. Through an understanding of how plants function at the molecular level, the conversion of solar energy into yield may be improved. For example, this could be achieved by modifying the plant structure so that it harvests solar energy more efficiently or by modifying biochemical pathways to direct more products of photosynthesis to seeds.

2. Food Quality

Almost as important as crop yield is the quality of the product. Some of the major subsistence crops lack vitamins and minerals important in the human diet. For instance, rice lacks provitamin A and iron which can be supplemented by other foods. If a rice diet lacks these other food supplements, vitamin A deficiency can lead to blindness in children and iron deficiency to anaemia in children and pregnant women. These problems are widespread in some developing countries. The development of GM "golden rice" (Box 1.8) shows promise in helping to mitigate vitamin A deficiency. This is an example of biofortification and demonstrates how GM can be used to improve food quality.

Another target for GM is the possibility of reducing or eliminating certain food and non-food allergens. Allergies are caused by the body reacting to certain proteins (allergens). About 1–2% of the adult population has life-threatening allergies to proteins in food products such as cow's milk, cereals and nuts or to proteins in pollen from certain plants. These proteins can be either eliminated or modified by GM so that they do not cause problems. (But see Chapter 3 on risk assessment for human health.)

3. Post-Harvest Protection and Processing

The division between food producers and food consumers, and urbanization of the population, means that food has to be stored and distributed. It is estimated that post-harvest losses range from 10 to 40%, depending on the crop and country, due to a range of biotic and processing factors (harvesting, milling, etc.). GM offers the potential to control losses due to biotic factors such as fungal or insect damage, and to limit premature ripening during storage and transport. An example of controlling premature ripening is Flavr Savr tomato (Box 1.9 and Case Study 5, Appendix D), which uses a delayed ripening strategy applicable to many fruits.

BOX 1.8

GOLDEN RICE

At the beginning of the 21st century, 124 million people, in 118 countries in Africa and South-East Asia, were estimated to be affected by vitamin A deficiency (VAD). VAD is responsible for 1–2 million deaths, 500,000 cases of irreversible blindness and millions of cases of xerophthalmia annually. Children and pregnant women are at highest risk. Vitamin A is supplemented orally and by injection in areas where the diet is deficient in vitamin A. Because many children in countries where there is a dietary deficiency in vitamin A rely on rice as a staple food, the genetic modification to make rice produce pro-vitamin A is seen as a simple and less expensive alternative to vitamin supplements or an increase in the consumption of green vegetables or animal products.

Golden rice is a variety of rice genetically modified to biosynthesize beta-carotene, a precursor of pro-vitamin A in the edible parts of rice. Figure 1 shows a simplified overview of the carotenoid biosynthesis pathway in golden rice.

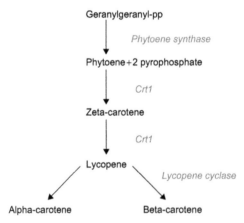

FIGURE 1   Biochemical pathway for the synthesis of beta-carotene. The enzymes involved are shown in red (see colour section).

The enzymes expressed in the endosperm of golden rice, shown in red, catalyse the biosyntheis of beta-carotene from geranylgeranyl diphosphate. Beta-carotene is assumed to be converted to retinal and subsequently retinol (vitamin A) in the animal gut.

Golden rice was created by Ingo Potrykus of the Institute of Plant Sciences at the Swiss Federal Institute of Technology and Peter Beyer of the University of Freiburg who transformed rice with two beta-carotene

(cont'd)

BOX 1.8   (*cont'd*)

biosynthesis genes: (1) *psy* (phytoene synthase) from daffodil (*Narcissus pseudonarcissus*) and (2) *crt1* from the soil bacterium *Erwinia uredovora*. The *psy* and *crt1* genes were transformed into the rice nuclear genome and placed under the control of an endosperm-specific promoter, so that they are only expressed in the endosperm. The original golden rice was called SGR1, and under greenhouse conditions it produced 1.6µg/g of carotene. In 2005, a new variety called *Golden Rice 2* was announced which produces up to 23 times more beta-carotene than the original variety of golden rice. Both varieties have been crossed with local rice cultivars and field tested but neither is currently available for human consumption. Although golden rice was developed as a humanitarian tool, it has met with significant opposition from environmental and anti-globalization activists.

There were 72 intellectual property rights belonging to 32 companies in the technology used to produce golden rice. Potrykus has spearheaded an effort to have golden rice distributed for free to subsistence farmers. Free licences, so-called humanitarian use licences were granted quickly due to the positive publicity that golden rice received. Golden rice was said to be the first genetically modified crop that was inarguably beneficial, and thus met with widespread approval. The group also had to define the cut-off between humanitarian and commercial use which was set at US$10,000. Therefore, as long as a farmer or subsequent user of golden rice genetics does not make more than $10,000 per year, no royalties need be paid for commercial use. There is no fee for the humanitarian use of golden rice, and farmers are permitted to keep and replant seed.

---

4. Pharmaceutical and Industrial Products

The plant can be considered as a "bioreactor or biofactory making use of a free source of energy, solar energy". Thus far, humans have used this system mainly for food production. Advances in GM technology open up a wide range of other possibilities which could lead to further opportunities for rural economies. Two examples of these opportunities are the potential to produce pharmaceuticals and industrial products in plants. Some examples of products that are in the pipeline for release as GM crops are given below. However, there are associated biosafety issues which will be discussed in subsequent chapters.

Vaccines can be produced by engineering plants to express antibodies or epitopes (protein sequences that elicit the immune response in humans and other animals); alternatively, plant viruses have been engineered to produce epitopes. One advantage of this approach over the current one of producing vaccines in animal cells is that plants do not have human-infecting pathogens that might be present in animal cells.

BOX 1.9

FLAVR SAVR TOMATO

The ripening and softening of tomatoes involves the enzyme polygalacturonidase (PG) which digests the cell walls. Unmodified tomatoes are picked before fully ripened and artificially ripened using ethylene gas which acts as a plant hormone. Picking the fruit while unripe allows easier handling and extended shelf-life but reduces the flavour.

To enable tomatoes to be picked when ripe and to prevent softening and rotting, tomato plants were genetically modified with an antisense RNA to the messenger RNA for PG reducing the expression of the enzyme by RNA interference. These tomatoes could be allowed to ripen on the vine, without compromising their shelf-life. The intended effect was to slow down the softening of the tomatoes, so that vine-ripe fruits could be harvested like green tomatoes without greater damage to the tomato itself.

This genetic modification was done in two centres: by Calgene in California and by Zeneca in the UK. The Calgene tomato was called Flavr Savr, which was the first commercially grown GM food to be granted a licence for human consumption. It was first sold in 1994, and was only available for a few years before production ceased. The Flavr Savr disappointed researchers as the antisensed PG gene had a positive effect on shelf-life, but not on the fruit's firmness, so the tomatoes still had to be harvested like any other unmodified vine-ripe tomatoes. The initial variety of tomato Calgene was also considered by farmers to be inferior. These and various other production and commercial problems led to its failure.

Zeneca were involved in an intellectual property dispute with Calgene which was resolved in their modified tomato being used to produce a tomato paste. The higher pulp content of this GM tomato enables more efficient processing of the thick pastes and ketchups preferred by consumers.

These early attempts at genetically modified food raise several interesting points:

- They were the first foods released for human consumption and were generally accepted by the public.
- The selection of the tomato variety was a major factor in Flavr Savr failing commercially.
- This was the first intellectual property dispute over a GM crop.

Plants are being engineered to produce industrial raw materials and industrial enzymes. For instance, plant fatty acids are being modified so that they can be used as renewable hydrocarbons such as biodiesel, and plant starches are being modified for various uses such as production of degradable plastics.

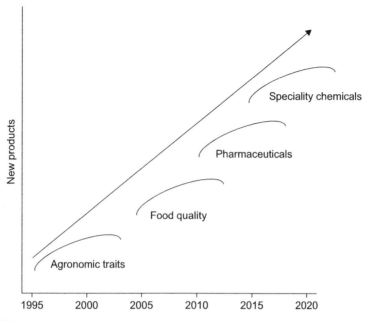

FIGURE 1.9 Anticipated initial delivery times of new products of biotechnology – further examples of each product type will continue to be developed after the initial time. Adapted from Fraley (1994).

An indicative timeline for the release of categories of GM products is shown in Fig. 1.9. It should be noted that this is based on products in the pipeline and may be affected by the time to obtain regulatory approvals.

V. COMPARISON OF CONVENTIONAL BREEDING AND GM

Table 1.5 shows a comparison of the conventional breeding and GM approaches to producing a new variety.

This table and Fig. 1.4 emphasize that the main difference between the two approaches is the source and means of introduction of new traits; otherwise, the two technologies are very similar. It should also be remembered that the sources of variation and new traits used in conventional breeding include those produced by mutagenesis (see Box 1.3) where there may be unrecognized changes to the plant genome.

VI. BIOSAFETY ISSUES

As with many new technologies, GM technology has raised a wide range of concerns.

TABLE 1.5   Comparison of conventional breeding and GM.

	Recombinant technology	Conventional breeding
Gene pool	Unlimited.	Usually limited to relatives within species (except with wide crosses).
Number of inserted genes	Usually, one or few known genes.	Many blocks of genes, usually of unknown identity. Genes of interest are transferred together with clusters of tightly linked genes (genetic drag).
Gene expression	Transgene expression can be monitored during subsequent breeding cycles.	The expression of individual genes introduced by crossing cannot be monitored.
	Broad host range promoters able to be switched on in many different plant species.	Narrow host range promoters.
Location of genes	Random into the recipient genome.	Usually, genes remain in locations in which they evolved.
Speed and accuracy	Relatively fast. Detailed knowledge at molecular level of inserted gene(s).	Relatively slow requiring many back-cross generations to give desired phenotype. Genes of both parents are mixed with little molecular knowledge on final outcome.

The first GM experiment to be published was in October 1972 in a paper describing the insertion of bacterial virus genes into an animal virus (SV40) DNA. This led to concerns from scientists about the potential risks of such experiments to human health being published in 1973 and 1974, and to scientists agreeing at the Gordon Conference on Nucleic Acids in 1973 to voluntarily suspend many such experiments. The follow-up Asilomar Conference in California in February 1975 which was attended by scientists, lawyers, the media and government officials discussed the potential risks of this technology (recombinant DNA technology). They initially endorsed a moratorium and concluded that such experiments could proceed once strict guidelines had been drawn up by the National Institutes of Health (NIH). Thus, the need for regulations was instigated by the scientific community.

The US government put the NIH guidelines into law (the recombinant DNA regulations) in 1977 which have subsequently been amended to take account of rapid advances in the technology (see Chapter 6, Section I.C). This area of law raises a very interesting point. All areas of law, except one, are *reactive* to a risk been shown; the one exception is the biosafety laws which are *proactive*, being triggered before a risk has been demonstrated.

Since the 1970s, GM biosafety regulation laws have been passed in most industrialized countries and in an increasing number of developing countries. The European Union (EU) countries have regulations governed by the EU directives and regulations (the Environmental Directive (2001/8), the Novel Food Regulation (258/97), the Genetically Modified Food and Feed Regulation (1829/2003) and the Traceability and Labelling Regulation (1830/2003)).

As well as the national and regional regulations, a set of international agreements and guidelines regulating biotechnology has been developed. These include the World Trade Organization (on trade), the World Health Organization and Food and Agriculture Organization Codex Alimentarius (food safety), the International Treaty on Plant Genetic Resources for Food and Agriculture and the Cartagena Protocol of the Convention on Biodiversity (CBD). The CBD deals wholly with genetic modified organisms (termed living modified organisms (LMOs) in the Convention) and the Cartagena Protocol provides an international regulatory framework for the transboundary movement of LMOs aimed primarily at environmental protection. The Cartagena Protocol drives the adoption of regulatory structures by developing countries because, if they are signatories, they cannot trade in GM products unless they have an acceptable regulatory structure. These international agreements are discussed in detail in Chapter 6.

There are three areas of concern about the release of GM crops: possible risk to the health of humans and other animals, possible risk to the environment and socio-economic and ethical issues. The procedures for assessment and management of these potential risks are described in subsequent chapters. Socio-economic and ethical issues are dealt with separately (in Chapter 5).

A. Human and Animal Health

Concerns about human and other animal health revolve around the possibility that the introduction of one or more genes from completely unrelated organisms might produce toxins or allergens in the product. These effects might be directly from the gene itself or indirectly from the process of insertion of the gene(s). The potential hazards and risks associated with these concerns are discussed in detail in Chapter 3.

B. Environment

Environmental concerns focus on possible negative impacts that transgenic plants might have on the environment. There are two relevant environmental components: that within the agricultural system and the "natural" one outside the agricultural system. Within the agricultural system, the potential risks include the transgenic plant becoming a weed, the spread of the trait to weeds within the crop, possible effects on non-target organisms, spread of the transgene to non-GM crops, and the impact of any changes in agricultural practices. In the "natural" system concerns are that spread of the trait could adversely affect that ecosystem or that the transgene can spread into wild relatives of the crop in centres of origin and diversity. Chapter 4 describes how these potential risks to the environment are assessed.

C. Baseline for Risk Assessment

There have to be baselines for risk assessment against which any potential risk can be assessed. The generally accepted baselines are the non-GM agronomic and food safety systems. The non-GM systems include the processes of plant breeding described above and the inputs normally used to control biotic and abiotic constraints. The food safety systems are those that are used for novel crop products. These baselines for comparison are discussed in more detail in Chapters 2, 3 and 4.

References

Fraley, R.T. (1994). The contribution of plant biotechnology to agriculture in the coming decades. In *Biosafety for Sustainable Agriculture* (A.F. Krattiger and A. Rosemarin, eds). The International Service for the Acquisition of Agri-biotech Applications (ISAAA), Ithaca, USA and the Stockholm Environment Institute (SEI), Stockholm, Sweden, pp. 3–28.

Hull, R. (2009). *Comparative Plant Virology*. Elsevier Academic Press, Amsterdam.

James, C. (2007). *Global status of commercialized biotech/GM crops: 2007. ISAAA Brief* No. 37. ISAAA, Ithaca, NY.

James, C. (2008). *Global Status of Commercialized Biotech/GM Crops: 2008.* ISAAA Brief No. 39. ISAAA, Ithaca, NY.

Mawer, S. (2006). *Gregor Mendel: Planting the Seeds of Genetics.* Abrams, New York. PG Economics Ltd (2008). Biotech crops: the real impacts 1996–2006 – yields. http://www.pgeconomics.co.uk.

PG Economics Ltd (2008). Biotech crops: the real impacts 1996–2006 – yields. http://www.pgeconomics.co.uk

Pray, C.E., Huang, J.K., Hu, R.F. and Rozelle, S. (2002). Five years of Bt cotton in China – the benefits continue. *Plant J.* **31**, 423–430.

Ülker, B., Li, Y., Rosso, M.G., Longemann, E., Somssich, I.E. and Weisshaar, B. (2008). T-DNA-mediated transfer of *Agrobacterium tumefaciens* chromosomal DNA into plants. *Nature Biotechnol.* **26**, 1015–1017.

Vavilov, N.I. (1935). *Theoretical Basis for Plant Breeding*, Vol. 1. Moscow. Origin and geography of cultivated plants. In *The Phytogeographical Basis for Plant Breeding* (D. Love, transl.). Cambridge University Press, Cambridge, UK, pp. 316–366.

Further Reading

Chrispeels, M.J. and Sadava, D.E. (1994). *Plants, Genes and Agriculture*. Jones and Bartlett Publishers, Boston, MA.

Principles of Risk Assessment

ABSTRACT

Risk assessment is based on a set of basic principles and follows a well-defined structure. This chapter describes the basis of risk assessment, the typical structure of the risk assessment process, and how it is practically applied at different points during the development and release of GM crops.

OUTLINE

Genetically Modified Plants 33

I. INTRODUCTION

A. Preamble

Humans are confronted with risk on a daily basis and accept or reject it on the basis of personal or aggregate social experiences in dealing with the source of risk. How one deals with risk depends on how risk is perceived and this is, in turn, determined by numerous factors. We shall address the issues related to risk perception in Chapter 5. Governments and industry, likewise, deal with risk on the basis of experiences that have arisen from systemic failures. In time, these cumulative experiences have been codified in the form of operational rules, codes of practice and legislation. Although traditionally risk has been viewed from the perspective of perceived harm to human health from a particular action, more recently, the perception of risk has been broadened to include environmental considerations. In the case of agriculture, the application of conventional farm technologies largely ignored environmental impacts until relatively recently when legislators and civil society recognized that the potential impacts of agricultural technologies and practices extend beyond the farming systems in which they are applied and have the potential to affect ecosystems and ecosystem processes. In many parts of the world, the application of transgenic technologies in agriculture triggered new regulations requiring the safety assessment of genetically modified organisms (GMOs) and derivative products prior to their environmental release and commercialization. The assumption is that, as the GMO genotype contains genes from taxonomically distant species, there may be risks over and above those presented by the conventional methods of crop improvement. The main differences between conventional breeding and recombinant technologies were described in Chapter 1, section V.

The question often arises as to whether transgenic technologies present new categories of risk when compared with those that may arise from conventional agricultural practices. As will be discussed in Chapter 6, in most countries the regulatory structures address specifically GMOs; only in Canada do the regulations focus on "novel plant products" which encompasses plants produced by conventional breeding as well as those arising from GM technology.

B. Potential Risks from GMOs

It is difficult to make broad generalizations about the potential risks associated with a GMO because the different types of introduced traits pose different types of risks which can arise from the phenotype (see list of commonly used terms in Appendix A) or from the use of the product. For example, transgenes conferring pesticidal traits present different potential risks from transgenes that are intended to directly increase crop yield. Potential harm to human health or the environment may be caused directly by a specific gene product, or by the whole transgenic plant due to unanticipated changes in its phenotype (pleiotropic effects, see Appendix A), or indirectly from the escape of transgenes and spread among plant populations in cultivated or wild ecosystems through pollen flow or seed dispersal. Some indicative risks are shown in Table 2.1.

C. Basic Definitions

In relation to GMOs, it is assumed that the novel trait(s) or the actions involving the transgenic organism itself may cause harm to human health and the environment. GM hazard is, therefore, defined as an intrinsic property of the GMO or activity involving the GMO and the circumstances pertaining to its use that may cause harm. However, harm cannot be manifested unless susceptible biota are exposed to the source of harm (i.e., the causative agent). Risk is, therefore, the likelihood of harm that may occur under realistic conditions of exposure.

Risk assessment refers to a structured process by means of which we attempt to link cause and effect. A formal definition of risk assessment is the "process of evaluation including the identification of the attendant uncertainties, of the likelihood and severity of an adverse effect(s)/ event(s) occurring to man or the environment following exposure under defined conditions to a risk source(s)" (European Commission, 2000).

TABLE 2.1   Indicative direct and indirect risks arising from the GM phenotype, the introduced trait or changes in agricultural practice.

Risk source	Potential risks	Mechanism
GM phenotype	Evolution of increased weediness (Direct)	Sexual transfer of crop alleles to wild relatives; seed dispersal
GM phenotype	Loss of biodiversity in the wild (Indirect)	Extinction by hybridization. Indirectly, from the intensification of agriculture
GM trait	Harm to non-target organisms (Direct)	Toxicity. Starvation through reduction of food resources
GM trait	Evolution of resistance in the targeted pathogen, pest or weed population (Direct)	Selection pressure from transgene products (e.g., Bt toxin) or application of agricultural input (e.g., herbicide)
Change in agricultural practice	Loss of agricultural biodiversity (Indirect)	Increased use of chemical inputs

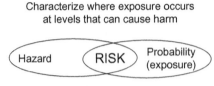

FIGURE 2.1   Basic risk assessment paradigm.

This process culminates in the description of risk as a function of the probability of harm given that exposure occurs. For environmental releases of GMOs, quantification of risk on the basis of probabilities is not possible for reasons that are analysed in depth later in this Chapter.

$$Risk = p(Harm) \times p(Exposure)$$

This is the basic risk assessment paradigm which is illustrated in Fig. 2.1 (see also section II.B.4.a below). Thus, not all hazards produce a significant risk – risk is dependent upon the probability of exposure to the hazard but there is no such thing as no risk. In this book, "probability of occurrence" will be used interchangeably with "exposure probability" or simply "exposure".

There are other definitions of risk assessment but, regardless of definitions, in practical terms risk assessment of GMOs attempts to answer the following questions:

- Can the GMO or its offspring cause adverse effects to humans, animals and/or the environment?
- If the transgene(s) is transferred to other organisms, could the latter be the cause of adverse effects?
- Are there effective measures for risk mitigation and/or management?

Answers to these questions determine to a large extent decisions as to how to manage risk and whether the environmental release of a GMO should proceed. The way in which risk hypotheses and exposure pathways are formulated, information is gathered and analysed, and decisions regarding the relative environmental or human and animal safety of a GMO are made constitute the methodological framework of risk assessment.

II. RISK ASSESSMENT METHODOLOGY

A. Risk Assessment Frameworks

There are three basic approaches to risk assessment: (1) empirical, which is based on scientific evidence and real world experience; (2) model based, which relies on predictive models; and (3) qualitative, which uses judgemental reasoning to draw approximate conclusions. Early efforts to develop GMO risk assessment frameworks were based on safety assessment rules and procedures pertaining to chemicals as the latter were developed much earlier and had a long regulatory history. For example, as early as 1993, the National Research Council (NRC) in the USA (National Research Council, 1993) developed a framework for health risk assessment consisting of four distinct stages, namely:

- Hazard identification: The determination of whether a particular chemical (stress agent or stressor) is causally linked to particular ecological effects;
- Exposure assessment: The determination of the extent of exposure before or after application of regulatory controls;
- Dose–response assessment: The determination of the relationship between the magnitude of exposure and the probability of occurrence of the effect in question;
- Risk characterization: The description of the nature and magnitude of risk, including associated uncertainty.

However, this framework was considered unsuitable for environmental risk assessment (ERA) because it required dose–response assessment which is difficult or impossible given the ability of living organisms to reproduce and disperse. There have been various attempts to overcome this limitation, two of which (both empirical) have led to the current approach.

The US Environmental Protection Agency (EPA) proposed a more qualitative framework (US EPA, 1998) comprising the following sequential stages:

- Problem formulation involving planning, information collection and selection of assessment endpoints, as well as the preparation of a conceptual risk model;
- Risk analysis involving acquisition and integration of data into existing datasets as well as characterization of exposure and effects (ecological responses);
- Risk characterization involving estimation of risk, evaluation of exposure and description of risk;
- Risk management involving practices to mitigate or manage risks.

The European Union adopted a more formalistic risk analysis framework consisting of three separate but interconnected components: risk assessment, risk management and risk communication. The risk assessment component involves hazard identification, hazard characterization, exposure assessment and risk characterization (EU, 2006). Although the terms used in this framework are somewhat different from those used by EPA, in practice they overlap. For example, risk hypothesis implies hazard identification and risk analysis encompasses hazard characterization and exposure assessment. Furthermore, both risk assessment frameworks are iterative. At each stage of the assessment, the information gathered and its interpretation may require additional data and/or risk management measures to be integrated into the model or even a review of the assumptions made in the entire risk assessment process. Frequently used terms and definitions in risk assessment are given in Box 2.1. There are no universally agreed definitions of terms and occasionally terms are used interchangeably.

BOX 2.1

DEFINITIONS

"Risk" is a measure of the likelihood and consequence of an undesired event. The consequences of an undesired event are usually expressed in terms of impact on human, economic or environmental values (Hayes, 2003).

"Risk assessment" is a process of evaluation including the identification of the attendant uncertainties, of the likelihood and severity of an adverse effect(s)/event(s) occurring to man or the environment following exposure under defined conditions to a risk source(s) (European Commission, 2000).

"Hazard" is the propensity for a substance or activity to cause harm. Hazard is a function of the intrinsic properties of the substance or activity, and the circumstances surrounding its use or implementation.

"Hazard identification" is defined as the identification of a risk source(s) capable of causing adverse effect(s)/event(s) to humans and/or the environment, together with a qualitative description of these effect(s)/event(s) (European Commission, 2000).

"Assessment endpoint" is an explicit expression of the environmental value that is to be protected and such values must be operationally defined by an ecological entity and its attributes (US EPA, 1998).

"Exposure assessment" is concerned with the likely actual levels and duration of exposure to the risk source of humans and the environment. An exposure assessment characterizes the nature and size of human populations and/or ecological communities exposed to an emission source and the magnitude, frequency and duration of that exposure (European Commission, 2000).

"Hazard characterization" aims at quantifying the relationship between exposure and effect(s) ("dose response"). In practice, the quantification of the interaction with or the uptake of each component of the risk source with the susceptible biota ("exposure dose" and "absorbed dose", respectively) is extremely difficult in all but the simplest cases of pathogenicity or toxicological assessment and simply not possible in ecological risk assessment (European Commission, 2000).

(cont'd)

BOX 2.1   (cont'd)

"Risk characterization" is the quantitative or semi-quantitative estimate, including attendant uncertainties, of the probability of occurrence and severity of adverse effect(s)/event(s) in a given population under defined conditions based on hazard identification, hazard characterization and exposure assessment (European Commission, 2000).

"Risk management" is the process of weighing policy alternatives in consultation with all interested parties, considering risk assessment and other factors relevant for the protection of consumers' health and for the promotion of fair trade practices as well as, if necessary, selecting appropriate prevention and control options (http://www.fao.org/docrep/003/x9602e/x9602e06.htm#).

"Risk communication" means the interactive exchange of information and opinions throughout the risk analysis process as regards hazards and risks, risk-related factors and risk perceptions, among risk assessors, risk managers, consumers, feed and food businesses, the academic community and other interested parties, including the explanation of risk assessment findings and the basis of risk management decisions (http://www.fao.org/docrep/003/x9602e/x9602e06.htm#).

The iterative nature of risk assessment and risk management is depicted in a flow chart (see Fig. 2.2), which will be used throughout this book.

Figure 2.2 shows that the accumulation of information for the decision-making process comprises two major, interlinked elements, namely, risk assessment and risk management, each of which involves several procedural stages, which are described in more detail below. As noted above, there are interactions and reiterations between the various stages so that the best and most practical solution to any potential problem can be reached.

It is important to recognize that this process is scientifically based but it does have some limitations, including ignorance of hazard endpoints and overall scientific uncertainty because nothing can be described as "black or white" (see Box 2.2). Furthermore, there are systemic limitations such as difficulties in addressing the potential for indirect, cumulative and synergistic ecological effects arising from GMO introductions, in comparing the significance of risks and uncertainties in different agricultural contexts and operational environments, and in accurately evaluating ecological risks arising from large-scale or commercial applications from small-scale field trials. Notwithstanding these limitations, the process should be viewed in the context of the release of non-GM crops for which there is also the potential for environmental problems to arise and for which there is no formal risk assessment.

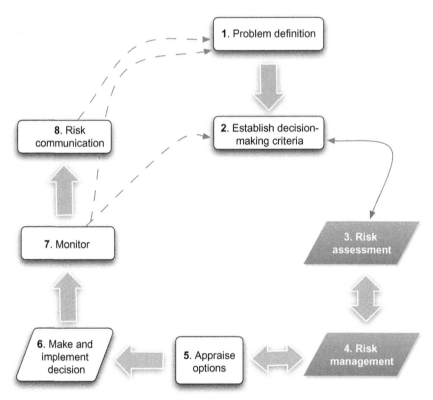

FIGURE 2.2 **1. Scope of the assessment** – Field trial; Food and feed safety; Environmental release; Commercial release; Import. **2. Establish decision-making criteria** – What are the regulatory requirements? Set up rules for decision making. **3. Risk assessment** – Identify risk source(s) and susceptible biota; Exposure scenarios; Assessment endpoints; Estimate magnitude of risk. **4. Define the risk management plan** – Consider alternative risk management options; Consider monitoring requirements. **5. Appraise risk management and monitoring options** – Evaluate options in terms of feasibility and cost effectiveness. **6. Decision making** – Were the decision-making criteria adequate? Has the problem been defined correctly? **7. Implementation of the monitoring plan** – Does the information gathered require amendment of the plan or redefinition of the problem? **8. Communicate risk** – Establish criteria for stakeholder participation; Actively engage different stakeholder groups; Do stakeholder inputs require redefinition of the problem? (see colour section.)

B. The Different Stages of Risk Assessment

1. Scope of the Release

This determines the extent to which hazards should be identified and, by implication, the extent of risk assessment. For example, in the case of contained applications of GMOs (see below), there is no need to identify potential environmental hazards and the main objective of the risk assessment process is to effectively avoid affecting the health of those working with it and the inadvertent escape of the GMO into the environment. In the case of environmental releases, the location and timing of the release and the size of the area covered by the GM plants are important considerations in the identification of risks. Similarly, the risk assessment criteria for the importation of GMOs are different from those that pertain to the development and release of local GM crop varieties.

2. Hazard Analysis

a. **Hazard Identification**   Hazard identification is the first stage of risk assessment (Fig. 2.2) and involves the identification of the source of risk(s). It is intended to identify potential impacts on human health and the environment (see Chapters 3 and 4, respectively, for detailed examples). In many cases, the potential impacts can be complex because of the interactions between genes, the GMO and the environment. The introduction of transgenic crops into the environment and/or the consumption of food and feed derived from such crops may give rise to concerns regarding the potential toxicity of the gene products, the consequences of outcrossing (see Appendix A) of the transgene(s), and possible evolutionary effects on target populations. Potential hazards are directly related to the input trait, the parent plant and the receiving environment. In this latter case, it also involves the description of the potential harmful effects at the species, population and ecosystem level and to what extent the occurrence of a particular hazard depends on the receiving environment and susceptible biota. This stage of the risk assessment aims at identifying the possible risk sources and biologically meaningful variables (endpoints) that can be measured as a test of a given risk hypothesis.

b. **Assessment Endpoint Identification**   This involves the identification of biologically meaningful variables that can be measured, permitting quantification of the potential harm. The identification of assessment endpoints is essential in proving a particular risk hypothesis and depends both on the identified hazard and the route of exposure of the susceptible organisms. Mortality levels (e.g., LD_{50}) could be one such endpoint in the case of toxicological assessments. Other examples of endpoints are fitness measures (e.g., seed set, survival, etc.) and population levels of important non-target organisms (see Table 2.2 for examples).

3. Exposure Assessment

Hazard analysis and exposure assessment are interlinked. Having identified the potential risk source(s) and the sequence of events that may cause harm, the next stage of risk assessment is to determine the frequency and duration of exposure of target and non-target receptors to the risk source. Exposure assessment is useful in establishing a relationship between risk source and effect.

TABLE 2.2 Examples of exposure assessment and assessment endpoint identification.

Risk source	Exposure conditions	Affected organisms	Effect	Assessment endpoint
Pesticidal trait	Feeding on plant tissues where the trait is expressed	Target pest	Emergence of resistance to pesticidal trait	• Pest population abundance • Damage to crop
	Trait expressed in nectar or pollen	Non-target species (insects and herbivores)	Loss of agricultural biodiversity	• Species population abundance • Damage to crop
	Trait expressed in seed	Seed-feeding animals	Loss of agricultural biodiversity	• Species population abundance • Damage to crop • Trait dispersal
	Trait expressed in roots	Soil organisms	Loss of agricultural biodiversity Loss of productivity	• Species population abundance • Damage to crop • Changes in ecosystem function
Herbicidal trait	Trait expressed in pollen	Sexually compatible wild species or weeds (in-field)	Loss of productivity	• Increased weediness
GM plant	Volunteer seed (in-field)	Rotational crop	Loss of productivity	• Increased weediness
GM plant	Dispersal through pollen or seed (off-field)	• Sexually compatible wild relatives • Herbivorous or granivorous animals	Loss of biodiversity	• Species population abundance

a. Spread and Dispersal In the case of field trials, this includes consideration of:

- The possible escape routes of GMOs and of inserted traits from the release site;
- The subsequent fate of the GMO, e.g. its potential for weediness and invasiveness;
- The possibility and routes of transfer of the transgene(s) by pollen or seed, hybridization, and dispersal of hybrids into new habitats;
- Assessment of any fitness advantage conferred to the GMO and/or hybrids and of interactions with target and non-target organisms.

In the case of commercial releases, transgene dispersal is taken for granted and the focus of the assessment is on long-term establishment and possible environmental impacts, such as adverse interactions with non-target beneficial organisms.

b. Effects Analysis   This stage of risk assessment is concerned with the determination of adverse effects from exposure to the GMO or to products of its transgene(s) at the species, population and ecosystem level. This may require a three-level analysis:

- **LEVEL 1.** Direct effects on target and non-target organisms under controlled laboratory or growth room conditions. Target and non-target organisms are exposed to high levels of the GM plant. The objective is to quantify effects on target and non-target organisms in relation to known exposure levels.
- **LEVEL 2.** Trophic level effects by assessing the effects on organisms not directly exposed to the GM plant under controlled laboratory or growth room conditions. In this assessment organisms removed by one or two steps in the food chain are exposed to high levels of the GM plant. The objective is to quantify indirect effects on non-target organisms in relation to known exposure levels.
- **LEVEL 3.** Ecological effects in field trials simulating the cultivation of the GM plant. The objective is to determine effects of exposure of different biota to the GM crop by using appropriate non-GM comparators (crop and management conditions).

This three-level analysis is conducted in a progressive manner as shown in Fig. 2.3.

Effects analysis of long-term impacts can be very speculative in the case of GMOs with long lifecycles and where ecological interactions are complex, such as in centres of biodiversity. In cases where effects analysis is inconclusive, it should be linked to monitoring (see below).

i. Effects on human health   This requires assessment of adverse effects resulting from use as human food (see Chapter 3) and also from the interaction of the GMO and persons working at the site(s) of the GMO release, particularly in those cases where the GM crop is not intended for human or animal consumption.

ii. Effects on animal health   This requires assessment of adverse effects on animal health resulting from the consumption of the GM crop or any derivative products if they are intended to be used as animal feed.

iii. Ecological effects   This requires assessment of adverse effects on the environment resulting from impacts of the environmental release of the GM crop. Such effects could be produced through direct toxic effects of the GM plant or indirectly through impacts of the GM plant on the abiotic environment. These can be measured by changes in agronomic

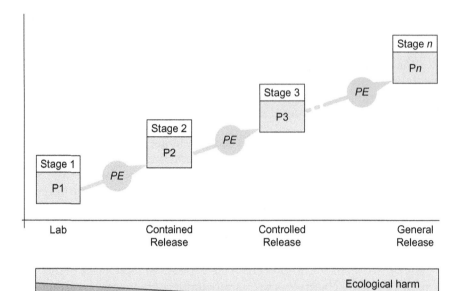

FIGURE 2.3 Schematic representation of the stages of ecological risk assessment for the different containment situations (laboratory to general release) starting with the formulation of the initial problem (P1) followed by a succession of Project Evaluations (PE) until the general release of the GM crop (P*n*). Each PE includes: identification of additional problems; modelling; problem analysis; conclusion; and decision. Concomitant to the progression of the containment situations, the level of ecological harm rises whereas the level of toxicological harm falls. The figure is modified from an original that was prepared by Dr Jeffrey D. Walt for the e-Biosafety International Training Network (http://binas.unido.org/moodle). (see colour section)

characteristics, alterations in plant or animal community structure and other means discussed further in Chapter 4.

iv. Effects on agricultural practices Changes in agricultural practices may be the cause of adverse or beneficial environmental impacts due to increases or decreases in the application of chemical inputs and/or water. Assessment of such indirect effects can be problematic due to the fact that agricultural practices even with the same GM crop are not uniform. This is discussed further in Chapter 4.

4. Risk Evaluation

a. Risk Characterization This stage of risk assessment involves the integration of data derived from the earlier stages of the assessment and the quantification of risk. The latter, as defined above in Section (I), is a function of the probability of the occurrence of an event that can cause harm. Thus, risk characterization is expressed in probabilistic terms, i.e.

$$Risk = Probability_{occurrence} \times Magnitude_{harm}$$

This is equivalent to the description given in section I.C.

b. *Data Integration* This is the description and compilation of data that relate to the exposure routes, such as different scenarios for gene flow, and linking these to the magnitude of predicted effects.

c. *Risk Quantification* Whereas quantification of risk is possible for toxicological risk assessment of foods and feed, it is not so in the case of ecological risk assessment for the reasons listed in Box 2.1.

5. Risk Management

Risk management is the decision-making stage of the risk assessment process and can be defined as the process of deciding what actions to take in response to an identified risk. Where the degree of uncertainty with regard to a particular risk is considered to be high, appropriate containment and mitigation measures are set or further research and monitoring are required. However, in some regulatory jurisdictions, risk management is taken to mean "a process distinct from risk assessment, of weighing policy alternatives in consultation with interested parties, considering risk assessment and other legitimate factors, and, if need be, selecting appropriate prevention and control options" (EFSA, 2006). This definition implies that decisions can be made based on criteria that are not strictly science based and to some extent decouple risk management for the earlier stages of risk assessment which are science based and concern the collection of data that permit the characterization of risk.

Risk management requires:

• Interaction between risk managers, risk assessors and other stakeholders;
• Transparency and legitimacy in the decision-making process; and
• Setting up appropriate monitoring mechanisms to evaluate the effectiveness of risk assessments and risk management responses.

6. Risk Communication

Effective interaction among different stakeholders depends on communicating risks in a way that takes into account scientific uncertainty, differences in scientific opinion regarding specific risks, and the asymmetric distribution of costs and benefits. These are issues that influence the perception of risk and will be dealt with in detail in Chapter 5.

III. DEALING WITH UNCERTAINTY: PRACTICAL ASPECTS OF RISK ASSESSMENT AND RISK MANAGEMENT

A. Information Requirements for Risk Assessment and Risk Management

The production of a GM plant for commercial release goes through five stages with different requirements for risk assessment and containment. The initial stages of designing and making the construct and transforming the plant takes place in the laboratory and growth chambers and the main concerns are potential risks to the health of the scientists and technicians. The putative transformed plants are usually then grown in glasshouses or screen houses to determine their phenotypic properties (see Appendix A) and to provide material for molecular analyses. Promising lines are then grown in controlled field releases to address various questions raised by biosafety committees (e.g., potential gene flow and potential for weediness) and also for the scientist to determine how the line responds to "natural" growing conditions. Lines that pass through these three stages are then evaluated for commercial release. There are important points to note from the flow chart in Fig. 2.4. Each stage is intended to generate information relevant to the overall development of the transgenic crop in terms of safety and performance, such as its phenotypic characteristics, agronomic performance and interactions with the environment outside the farming system.

The risk assessor needs to be able to make a distinction between what "needs to be known" from what "is nice to know". Data gathering and analysis have some inherent shortcomings and, in addition, there are a number of challenging systemic limitations due to uncertainties that render risk analysis an imprecise science and risk assessment procedures problematic (Box 2.2). These need to be addressed practically, if decisions are to be made and accepted.

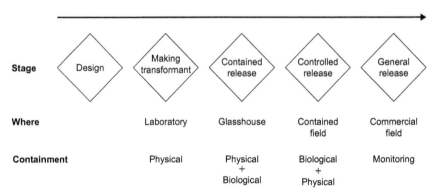

FIGURE 2.4   Stages for the production of a GM line for commercial production. The line of boxes shows the successive stages described in the text. The next line indicates where these stages occur and the bottom line shows the type of containment that is required.

BOX 2.2

SOME SCIENTIFIC UNCERTAINTIES

Systemic limitations include:

- Inability to build generic models due to wide variations in environmental and ecological conditions from location to location and difficulty in evaluating indirect, cumulative and synergistic ecological effects arising from GMO introductions;
- Dependence of environmental impacts on spacio-temporal factors. Cumulative effects at the population, community and ecosystem level are time dependent, and scale dependent, that is the larger the scale of the release the greater the probability of an adverse effects impact;
- Inability to identify all risk sources and all routes of exposure or co-exposure;
- Difficulty in evaluating ecological risks arising from commercial, large-scale applications by extrapolation from small-scale field trials.

Data limitations can arise from absence of specific data, poor data quality, possible unidentified inter- and intra-species variation. They can be broadly categorized as follows:

- Statistical: Even in cases where there exist quantitative data measuring correlation between harm and exposure to *a priori* identified hazards, these data may be prone to statistical error attributing correlation where none exists (significant number of false positive data, Type I Error) or failing to attribute correlation where one does exist (significant number of false negatives, Type II Error). Type II errors are more relevant in GMO risk assessment and, as such, should be the focus of risk assessment.
- Modelling: This arises from the fact that most models are reductionist in nature, thus giving inadequate representations of the real biological or ecological situations. For example, there is an inherent problem in developing sound risk hypothesis models for the causal chains of events giving rise to risk (exposure assessment) and/or collecting data that are relevant for the measurement of hazard endpoints. Nevertheless, models are still extremely useful in that they provide a methodological framework for the collection of relevant data.
- Ignorance: This arises either because it is difficult to identify hazards *a priori* or because there is no consensus on what constitutes a hazard or not. Historically, many industrial hazards have been recognized *a posteriori*, as, for example, the bioaccumulation of heavy metals and ozone depletion from chlorofluorocarbon emissions. In the case of GMOs, dealing with ignorance is an important issue.

The practicalities of the information requirements for making a risk assessment are shown in Appendix B, which is a generic outline for applicants for GM release permission listing the types of questions that might be asked by a regulatory authority and the forms of data that should be provided to address those questions, and Appendix C which justifies the questions.

The stage of GM product development and the objective of the experiment or the scope of the release determine the precise nature of the information that needs to be analysed, as well as the type of management procedures that need to be put in place.

- For many experimental situations, contained or controlled release is all that is required in order to decide whether a new trait has any realistic potential. For instance, for a gene, the product of which might confer drought resistance, the *in vivo* properties of this gene may be tested in plants which are easy to transform (e.g., *Arabidopsis thaliana* or tobacco) before attempting to transform more recalcitrant crop plant species.
- If the product is intended to go to the commercial release stage, it is important to consider all the biosafety implications of the construct(s) and the processes of transformation before the work starts. Producing a transformed elite crop variety line is a long process and it is essential to go through a risk assessment at the planning stage to identify any potential problems and take measures to mitigate them. An example is the use of particular antibiotic resistance genes used as selectable markers (see Table 1.2) which are considered to be undesirable in the final product (see Chapter 3).

B. Case-by-Case and Step-by-Step

The above considerations led to the development of the "case-by-case" and "step-by-step" GMO risk assessment approach. This was first laid down in a seminal report entitled "Recombinant DNA Safety Considerations" (OECD, 1986) which stated that "It is important to evaluate rDNA organisms for potential risk, prior to applications in agriculture and the environment. An independent review of potential risks should be conducted on a case-by-case basis prior to the application. Development of organisms for agricultural or environmental applications should be conducted in a stepwise fashion, moving, where appropriate, from the laboratory to the growth chamber and greenhouse, to limited field testing and finally, to large-scale field testing." Case-by-case is taken to mean "an individual review of a proposal against assessment criteria which are relevant to the particular proposal". As noted above, each stage in the development of GMOs requires different risk management measures in the form of containment, confinement and monitoring and is intended to generate data that can be useful in the prediction of potential risks.

Although this approach was born out of the relative inexperience with GMO releases at the time, it is still enshrined in most regulatory jurisdictions which mandate the preparation of comprehensive risk assessment and risk management plans on a per case basis. Below, we examine some practical risk assessment and management approaches for each stage of the development and commercialization of transgenic crops.

1. The Design Stage

Best practice refers to proactive GM crop design so that risks and, consequently, the regulatory oversight burden are reduced. Risk reduction can be achieved by any of the following approaches:

- Minimize extraneous DNA;
- Minimize transgene dispersal in the environment; and
- Minimize unnecessary transgene expression.

These approaches effectively reduce exposure to the risk source thereby minimizing the potential for an adverse effect (harm). If exposure to the risk source is shown to be close to zero, then the putative risk itself can be treated as close to zero. Methods aiming to achieve the above objectives have been developed or are currently under development. Methods that aim at minimizing transgene dispersal are often referred to as bioconfinement. Table 2.3 summarizes these methods and their advantages and shortcomings.

2. The Laboratory and Greenhouse Stages

Containment at these stages is primarily physical (e.g., disposal of plant material, controlling air flow and water flow) but with some biological containment (e.g., controlling pollen flow). During these stages of the development of GM plants, good laboratory practice is needed to ensure the containment of transgenic plant material within the laboratory. Good laboratory practices involve proper controls to avoid mixing transgenic and non-transgenic material, and the disposal of any unwanted material through treatment with chemicals, sterilization, autoclaving, or any other suitable method appropriate to the characteristics of the organism, the product and the process. It can be achieved by ensuring that personnel are adequately trained and, if required, the establishment of a local (institutional) biosafety committee (see Chapter 6) and effective mechanisms of consultation with non-scientific personnel (see OECD, 1992).

Greenhouses provide environmentally controlled areas in which the growth of GM plants can be monitored while ensuring zero or minimal exposure of GM plants to insects, pests and diseases. Greenhouses are constructed to afford different levels of containment depending on the

TABLE 2.3   Methods for minimizing transgene dispersal.

Purpose	Method	Comments
Removal of extraneous DNA	**Cotransformation** Available. Technically simple.	Cannot be used with non-meiotic plants or with plants with long generation time
	Excision of marker genes though site-specific recombination[*] Available.	Involves a recombinase enzyme. Risk assessment would need to conduct molecular analysis to confirm the excision of the marker, consider the toxicological or allergenic properties of the enzyme and potential unintended recombinase-mediated rearrangements.
	Transposon[*]-mediated repositioning of marker genes Available. Allows selectable marker-free plants to be identified in the segregating progeny.	Low frequency of excision of the selectable marker. Rigorous screening required to define the site of insertion, confirm the absence of unwanted sequences and rearrangements.
Reduce gene flow via pollen and seed	**Use of sterile triploids or interspecific hybrids** Available.	Applicable in few cases only. Not appropriate if seed production is desired.
	Use only male or only female vegetatively propagated plants Available.	Not appropriate if same species or compatible relatives can cross-pollinate with unisexual plants. Or, if seed production is desired.
	Seed sterility. V-GURTs[*] Under development.	Not appropriate in systems where growers save seed.
Reduce pollen flow	**Male sterility** Available.	Sterility can be lost in later genera-tions; not appropriate if seed produc-tion is desired unless other plants act as pollen sources.
	Chloroplast transformation Available.	Not possible for plants with paternal inheritance of chloroplast DNA. Chloroplast transformation can be achieved for a narrower range of transgenes.
	Cleistogamy[*] Under development.	Results in self-pollination but pollen is unlikely to escape from the flower. Requires manipulation of floral genes.
	Apomixis[*] Under development.	Apomictic plants have a fitness advantage and could become invasive. If an apomictic gene is linked to a transgene it may lead to extinction by swamping of sexual populations.

(Continued)

TABLE 2.3 *(Continued)*

Purpose	Method	Comments
Reduce gene flow between GM crop and crop relatives	**Tandem constructs to reduce fitness in hybrids and hybrid progeny** Under development.	Requires the transgene to be linked to a fitness-reducing trait detrimental to wild plants. The possibility of gene escape cannot be completely eliminated. Introgression into wild relatives may result in reduced fitness of the wild population.
	Chromosome location in allopolyploids* Under development.	Applicable only to allopolyploids and only if relative has non-homologous chromosomes.
Minimize transgene expression	**Chemically inducible promoters** Available.	Environmental changes may affect the reliability of these systems which needs to be demonstrated on a case-by-case basis.
	Tissue- and organ-specific promoters Available.	Greater promoter efficacy needed in many cases. Confines the transgenic trait but not the transgene.

Sources: National Research Council (NRC) (2004); Department for Environment, Food and Rural Affairs, UK (2000).
*Term defined in Glossary in Appendix A.

familiarity with the GM crop under development. For example, higher levels of containment are required for GM crops producing pharmaceutical or chemical compounds. The specifications of high containment level greenhouses would typically include controlled and filtered air flow and water flow, special sterilization and decontamination areas for personnel and materials, and monitoring systems for the accidental escape of transgenic organisms. Pollen containment is extremely difficult and can only be achieved in facilities constructed specially for this purpose. However, it is possible to achieve effective confinement (as distinct from containment) by a number of methods as discussed below.

3. Small-Scale Field Trials

Small-scale field trials are intended to monitor the agronomic performance of transgenic plants under conditions which ensure low or negligible risk. The level of risk is dependent on:

- The characteristics of the plant(s) including the introduced gene(s). These include the reproductive biology of the plant, characterization of the genetic construct, the mode of action and fate of toxic compounds (if any), and interactions with other species.
- The characteristics of the field in which the GM plant is grown and its surrounding environment. These include the prevailing ecological and

environmental conditions relative to safety, the size and boundaries of the experimental field including buffer zones, and specific flora and fauna in the surrounding environment that may be affected.

- The application of appropriate experimental procedures. These include a statement of the objectives of the field trial, precise description of the experimental design including the method of introduction of the plant, the planting density and treatment patterns, description of the data to be collected and analysed, significance of the results in the context of the objectives, and detailed description of monitoring and mitigation protocols.

Risk minimization can be achieved through the application of *good developmental practice* (GPD) (OECD, 1992). Table 2.4 shows the aims of, and means to achieve, GDP.

However, small-scale field experiments are not sensitive enough to detect anything but large effects. Low probability and low magnitude events may remain undetectable. Furthermore, high variability between one experimental replication and another may result from the complex interactions that occur in an ecosystem. Thus, although the scope of small-scale trials is limited to testing a small set of organisms over a short period of time, they do provide much valuable data applicable to general releases.

4. *Large-Scale Unconfined Releases: The Commercialization Stage*

During this stage, the transgenes cannot be recalled and so the Familiarity Approach is applied for risk assessment and, whenever the degree of uncertainty is high, the Precautionary Approach is invoked and post-commercialization monitoring applied to manage uncertainty. Whereas "familiarity" is based on previous experiences with GM and non-GM crops grown in specific environments, the Precautionary Approach and monitoring are complementary measures dealing with uncertainty.

C. The Familiarity Approach

The Familiarity Approach is intended to simplify risk assessment procedures by recourse to previous assessments, current scientific knowledge, and experience with field trials and commercial releases of GMOs. The concept of "familiarity" was born out of the need to focus the risk assessment process on those traits that produce novel phenotypes by drawing on previous knowledge and experience with the introduction of similar crops (GM and conventional ones) and traits into the environment. The concept was developed by OECD (1993) and is described as the knowledge and experience available for conducting a risk/safety analysis prior to scale-up of any new crop plant in a particular environment. Further, it states that risk assessment should be "based on the characteristics of the organism, the introduced trait, the environment into which the organism

TABLE 2.4 Examples of aims and means to achieve good developmental practice.

Aim: Internal or external reproductive isolation of the transgenic plant from sexually compatible plants outside the experimental field

Method	Examples
Internal isolation. Select plants with an intrinsic biological barrier to reproduction	Plants that fail to produce F1-zygotes due to cross incompatibility because of incompatibility of the reproductive organs of plant from different parents; produce F1-zygotes that are not viable or are weak (chromosomal or cytological incompatibility); the produced F1 hybrids are sterile (chromosomal or genic sterility); produce F2 plants that are weak or sterile (hybrid breakdown)
External isolation. Separate the transgenic plants from the surrounding environment	This can be achieved by spacial isolation preventing pollen transfer by wind or insects. This is a method used routinely by plant breeders to ensure seed purity; ecological isolation (not a complete barrier) whereby two populations are sexually isolated by adaptation to different habitats; phenological isolation (not a complete barrier) whereby the flowering two populations within the same area occur in different periods. It can also be achieved by selecting plants that have specialized pollinators. The organs of such plants have co-evolved with the pollinator and this co-evolution results in genetic isolation.

Aim: Confinement of the transgene(s) within the field

Method	Examples
Select plants with characteristics that limit their survival and dispersal potential	The manipulation of the experimental plants or the environment into which they are introduced by conventional technologies can result in effective genetic or reproductive isolation. In addition, it can be achieved by the methods described in Chapter 4
Disarm the plant from any vectors that present risk of injury	When transformation vectors are used (e.g. *A. tumefacience*), this can be achieved by ensuring the removal of vector sequences associated with pathogenicity
For pesticidal plants, ensure that the experimental conditions are such that contact with external gene transfer vectors (e.g. insects, birds, mammals) is minimal and/or mitigate gene flow by air or water	The degree of confinement will be determined from the amount of knowledge regarding the mode of action of the toxin and its effects on target and non-target organisms. Whenever necessary, additional mechanical barriers such as caging the plants or instituting strict controls for the material produced in the field and its safe disposal after termination.

is introduced, the interactions between these and the intended application. Knowledge of, and experience with, any or all of these provides familiarity which plays an important role." Moreover, "Familiarity is *not* synonymous with safety; rather, it means having enough information to be able to judge the safety of the introduction or to indicate ways of handling the risks." The concept is also embedded in the revised EU directive on the deliberate release into the environment of genetically modified organisms (2001/18/EC, Annex II) where it is stated that "information from releases of similar organisms and organisms with similar traits and their interaction with similar environments can assist in the environmental risk assessment" and that "a comparison of the characteristics of the GMOs with those of the non-modified organism under corresponding conditions of the release or use, will assist in identifying the particular potential adverse effects arising from the genetic modifications".

The effect of applying the concept of familiarity is to focus the risk analysis on the risks posed by the "novel" phenotypic trait and its interaction with the environment. The "familiarity" model argues that a GM crop plant should behave like an untransformed plant other than for the introduced trait, which introduces only a small amount of novel genetic information into the ecosystem. The familiarity concept is similar to the concept of substantial equivalence which is usually applied to food and feed and which is discussed in Chapter 3.

1. Comparative Assessment

The Familiarity Approach requires evaluation of the relative safety of a GM crop versus suitable conventionally bred comparators, with the latter forming the baseline of comparison (Fig. 2.5).

Depending on the potential hazard, various baseline or reference scenarios could be chosen. There are two general approaches for comparative

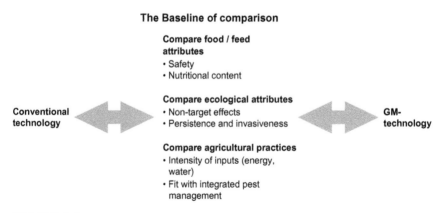

FIGURE 2.5   Characteristics of conventional crops and agricultural practices that are used as a baseline for assessment of GM releases.

risk assessment: a reductionist one that looks separately at the risks posed by the transgene and the host organism, and a holistic one that looks at the risks posed by the GM plant as a whole. The former approach reduces the GM plant into components, namely the untransformed plant and the inserted transgene(s), and then seeks appropriate comparators for the transgene product(s) to evaluate its relative safety and assesses separately the untransformed plant on the basis of current agricultural practices. If a crop with a similar phenotype can be identified, then it is used as an appropriate comparator. If it is shown to have a record of safe use, then the transgenic plant is deemed to be safe too. One such example would be transgenic glyphosate-tolerant soybean. According to this approach, given that a number of non-transgenic herbicide-tolerant soybean varieties are generally accepted as safe, the glyphosate-coding transgene would represent an incremental risk on which the risk assessment should be focused. In a second stage of the assessment, the whole transgenic plant is assessed in order to ensure that all potential hazards have been identified. This model implicitly assumes that the inserted transgene represents a very small amount of genetic change and as such its introduction into the environment would be likely to have no or small adverse ecological effects.

The holistic model considers the risks that may arise from the GM plant as a whole and in this regard, it would appear to be identical with stage two of the reductionist model. However, it differs from it in its basic philosophy according to which potential risks may arise from the transgene(s) in a plant and in a specific environment and not from the trait independently of the plant and the environment in which it is released.

The reductionist model has been challenged on two accounts. First, the implicit assumption that the introduced transgene invariably represents a small amount of genetic novelty, and therefore a small amount of risk, is not correct as empirical evidence has shown that both small and large amounts of genetic novelty introduced into the environment can have substantial environmental consequences and the consequences of biological novelty depend strongly on the specific environment. Second, the model introduces biases that may lead to overestimation or underestimation of risk. For example, if the two independent parts (the transgene and the untransformed host) are considered safe, that does not necessarily mean that the GM plant is safe and, vice versa, if the two parts are considered to be risky, the compounded risk is not necessarily the product of the two parts (National Research Council, 2002).

Regardless of any potential methodological or procedural limitations of the two models, comparative risk assessment is scientifically sound. Although ideally the best comparator for a transgenic crop plant would be its near-isogenic (see Appendix A) non-transformed parent, this may not always be possible. In this case, other baseline comparators may serve the purposes of risk assessment provided that the rationale for using them is appropriate. For example, in the case of transgenic Bt corn where no conventional Bt varieties exist, other insect-resistant corn varieties could be used as comparators. A more remote, but nevertheless appropriate, comparator would be the insecticide load in comparable agricultural fields growing Bt corn varieties and conventional ones.

D. The Precautionary Principle

The "Precautionary Principle" (Box 2.3) is a basic principle embedded in a wide range of international regulatory instruments.

BOX 2.3

THE PRECAUTIONARY PRINCIPLE AND PRECAUTIONARY APPROACH

The Precautionary Principle (*sensu stricto*)

The Precautionary Principle has been used in normal personal risk assessment for a very long time but the concept has been formalized and strongly developed since the 1980s. There are several versions of the Precautionary Principle including:

- "[When] potential adverse effects [of activities] are not fully understood, the activities should not proceed" (UN World Charter for Nature, 1982).
- "The Precautionary Principle applies where scientific evidence is insufficient, inconclusive or uncertain and preliminary scientific evaluation indicates that there are reasonable grounds for concern that the potentially dangerous effects on the environment, human, animal or plant health may be inconsistent with the high level of protection chosen by the EU" (EU communication on the Precautionary Principle, 2000, see European Commission (2003)).

The Precautionary Approach

- "In order to protect the environment, the precautionary approach should be widely applied by States according to their capabilities. Where there are threats of serious or irreversible environmental damage, the lack of full scientific certainty shall not be used as a reason for postponing cost effective measures to prevent environmental degradation" (Rio Declaration on Environment and Development, 1992).
- "lack of scientific knowledge or scientific consensus should not necessarily be interpreted as indicating a particular level of risk, and absence of risk, or an acceptable risk" (Annex III of the Cartagena Protocol on Biosafety, 2000).

The principle is used to address uncertainties in GM risk assessment and is implicit or explicit in a large number of international environmental agreements and treaties as a mechanism which facilitates dealing with human and animal health and environmental risks in the face of uncertainties. For example, the Cartagena Protocol on Biosafety of the Convention on Biological Diversity (CBD) is a major international legally binding agreement dealing with transgenic organisms (see Chapter 6) which states in Article 11.8 that "Lack of scientific certainty due to insufficient relevant scientific information and knowledge regarding the extent of potential adverse effects of a living modified organism on the conservation and sustainable use of biological activity in the Party of import, taking into account risks to human health, shall not prevent that Party from taking a decision, as appropriate, with regard to the import of the living modified organisms in question in order to avoid or minimize such potential adverse effects."

These Articles have an antecedent in the Rio Declaration of the United Nations of 1992 which forms part of the CBD and includes the following statement: "In order to protect the environment, the precautionary approach shall be widely applied by States according to their capabilities. Where there are threats of serious or irreversible damage, lack of full scientific certainty shall not be used as a reason for postponing cost-effective measures to prevent environmental degradation."

Most statements about the Precautionary Principle contain qualifying language and do not constitute strict definitions of the principle. Advocacy of the principle implies prevention of unacceptable human and environmental risks even in cases where there is absence of clear evidence of harm or the supporting evidence is speculative. Various interest groups have different interpretations of the Precautionary Principle influenced by their own agenda. This gives rise to considerable confusion as to the significance of any potential risk. The choice of the Precautionary Principle version influences the risk management decisions and outcomes including:

- Activation of the decision management process;
- Duration of the decision-making process;
- Efficacy of decisions; and
- The cost of risk management.

Some aspects of the Precautionary Principle need further explanation. In many cases, some interpretations of the Precautionary Principle *sensu stricto* remove the burden of proof from those who make (unjustifiable) claims and prevent scientific debate. As noted in Box 2.2 and Chapter 5, Section IV. B, scientific proof is rarely completely "black and white", and thus these interpretations of the Precautionary Principle can be, and have been, used in attempts to block the progress of GM (and other) technology. The threshold of its application depends on whether lack of knowledge regarding possible harm from a proposed activity/technology has been established. This raises the issue of who determines the level of knowledge and on what criteria. The principle also requires that the proponents of the technology carry the burden of proof of its safety. This is contrary to the traditional legal practice in which the burden of proof is on the side of the plaintiff. Furthermore, application of the principle means that imposition of a regulatory regime is possible without conclusive evidence of harm; it is sufficient to establish that harm is simply plausible. The principle is applied proactively to prevent harm before it occurs rather than reactively as has been the case with most risk management models that aim to eliminate harm once it has occurred. Last but not least, the principle has ethical implications as its application may result in opportunity costs. For example, should the principle be applied to mitigate putative environmental risks regardless of socio-economic objectives?

The above considerations complicate the interpretation of the Precautionary Principle and make its use as a risk management tool extremely difficult. These constraints on decision making have, instead, led to the use of the Precautionary Approach (Box 2.3) in which the need for precaution is balanced against other factors.

It is now widely accepted that the Precautionary Approach should be invoked when:

- There is good reason, based on empirical evidence or plausible causal hypothesis, to believe that harmful effects might occur, even if the likelihood of harm is remote; and
- A scientific evaluation of the consequences and likelihoods reveals such uncertainty that it is impossible to assess the risk with sufficient confidence to inform decision making.

These factors were involved in the decision by scientists in the early days of GM technology to voluntarily suspend recombinant DNA research (see Chapter 1, Section VI) until further discussion (the Asilomar Conference) had taken place and further information had been obtained.

The Precautionary Approach addresses the following questions (Nuffield Council on Bioethics, 2004):

- How does the use of a GM crop compare to other alternatives?
- What are the risks of the non-GM approach, which would constitute the option of "doing nothing"?
- In what respect are the risks posed by the introduction of a GM crop greater or less than those of the alternative system?
- Does the comparator system involve a higher level of benefits than the alternative system?

The European Commission, in a communication on the Precautionary Principle (European Commission, 2000), recommends that Precautionary Approach be:

- Proportional to the chosen level of protection;
- Non-discriminatory in its application;
- Consistent with other measures already taken;
- Based on an examination of the potential benefits and costs of action or lack of action (including, where appropriate and feasible, an economic cost/benefit analysis);
- Subject to review in the light of new scientific data; and
- Capable of assigning responsibility for producing the scientific evidence necessary for a more comprehensive risk assessment.

E. Monitoring

The shortcomings of small-scale field trials mentioned above, and uncertainties associated with commercial releases, may require that long-term *monitoring* mechanisms be put in place in order to detect any untoward environmental effects and record trends related to any predicted effects. An additional reason for setting up a GMO monitoring mechanism may be the need to reassure the public that concerns with regard to the safety of GM crops are adequately met and provide opportunities for public feedback. It can also identify any long-term beneficial effects that such crops may have on the environment.

Therefore the objectives of such monitoring potentially are two-fold, namely:

- The validation of earlier assumptions regarding risks and thus of the assessment protocols used; and
- The identification of any unanticipated adverse effects to human health and the environment resulting from large-scale releases of transgenic plants.

The first objective requires monitoring of specific adverse effects that have been identified in risk assessments. By implication, this type of monitoring should be carried out on a case-by-case basis through hypothesis-driven information gathering. Achieving the goals of the second objective is a different matter because no risk has been anticipated *a priori* and, therefore, no specific risk hypothesis can be tested. Instead, the goals of this objective can be achieved by general surveillance. Case-by-case monitoring does not exclude general surveillance, as the latter may still be required for the identification of unanticipated risks. In such cases, the demarcation line between case-by-case monitoring and general surveillance is blurred.

General surveillance is confronted with the following challenging points:

- How to identify unanticipated effects;
- Whether any unanticipated effects are detrimental; and
- Whether these unanticipated effects are due to the GM crop and not some other ecological perturbation.

There are many possible ecological interactions and it is a major challenge to identify whether an adverse effect identified during a commercial release of a GM crop is due to the crop itself and not some other factor. The issue here is not only what to monitor but also to identify the types of indicators that reflect changes in the items that have been selected for monitoring. These are essential elements in defining monitoring methods and approaches. The National Research Council (NRC) (2000) considered the criteria that would make an indicator appropriate for ecological monitoring, and these are summarized in Table 2.5.

TABLE 2.5   Criteria for indicators for ecological monitoring.

Indicator properties	Criteria
Importance	Relevance of the indicator with regard to information about perturbations in ecological processes
Conceptual basis	A well-understood conceptual model linking the indicator with ecosystem to which it is applied
Spatial/temporal scales	The indicator provides an adequate measure of short- and long-spatial and temporal ecological changes
Statistical properties	The indicator has statistical properties that are well understood and any changes in its values are interpretable and unambiguous
Data requirements	The type, amount and quality of data that need to be collected in order to detect trends in the indicator
Robustness	The indicator is insensitive to external ecological perturbations and technological changes affecting the monitoring protocols
Efficiency	No alternative indicators can be used for the same purpose at a lower cost

Adapted from National Research Council (2002) making reference to National Research Council (2000).

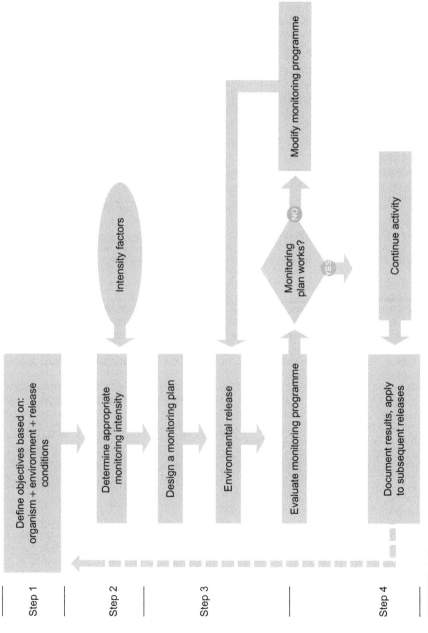

FIGURE 2.6 A basic approach to designing a monitoring programme.

Other important elements for the preparation of monitoring plans include:

- Plans for the introduction, stewardship and use of the GM crop;
- Proposals for the duration, frequency and area covered by the monitoring plan;
- Reference to existing monitoring systems and their respective merits and, whenever deemed necessary, proposals for the establishment of new ones;
- Arrangements for the collection and analysis of data, as well as methods for data archiving.

Furthermore, it has been argued (NRC, 2002) that the development of criteria for environmental monitoring programmes should be made through open multi-stakeholder consultations if monitoring plans are to gain the legitimacy that would make them publically acceptable.

A basic approach to designing a monitoring programme is shown in Fig. 2.6.

References

Annex III of the Cartagena Protocol on Biosafety (2000). http://www.cbd.int/biosafety/protocol.shtml

Department for Environment, Food and Rural Affairs, UK (2000). Advisory Committee on Releases to the Environment: Sub-group on Best Practice in GM Crop Design. Guidance on Principles of Best Practice in the Design of Genetically Modified Plants.
http://www.defra.gov.uk/Environment/acre/bestprac/guidance/index.htm.
http://www.defra.gov.uk/Environment/acre/bestprac/guidance/pdf/bestprac_plants_guidance.pdf.

EFSA (2006). Guidance Document of the Scientific Panel on Genetically Modified Organisms for the Risk Assessment of Genetically Modified Plants and Derived Food and Feed. *The EFSA Journal* **99**. p. 7 footnote and p. 12.
http://www.efsa.europa.eu:80/cs/BlobServer/Scientific_Document/gmo_guidance_gm_plants_en,0.pdf?ssbinary=true.

European Commission (2000). Report of the Scientific Steering Committee's Working Group on Harmonisation of Risk Assessment Procedures in the Scientific Committees Advising the European Commission in the Area of Human and Environmental Health – 26–27 October.
http://europa.eu.int/comm/food/fs/sc/ssc/out83_en.pdf.
Appendices: http://europa.eu.int/comm/food/fs/sc/ssc/out84_en.pdf

European Commission (2003). Communication from the Commission on the Precautionary Principle. COM (2000) 1 final. http://ec.europa.eu/environment/docum/20001_en.htm

Hayes, K.R. (2003). Robust methodologies for ecological risk assessment. Final report: Inductive hazard analysis of GMOs. CSIRO, Australia.

National Research Council (1993). Issues in risk assessment. In *National Research Council Committee on Risk Assessment Methodology. Board on Environmental Studies and Toxicology Commission on Life Sciences*. National Academy Press, Washington DC, 288pp.

National Research Council (2000). *Ecological Indicators for the Nation*. National Academy Press, Washington DC.

National Research Council (2002). Environmental Effects of Transgenic Plants: The Scope and Adequacy of Regulation. In *Committee on Environmental Impacts Associated with Commercialization of Transgenic Plants*. National Academy Press, Washington DC.

National Research Council (NRC) (2004). *Biological Confinement of Genetically Engineered Organisms*. The National Academies Press Washington, OK.

Nuffield Council on Bioethics (2004). *The Use of Genetically Modified Crops in Developing Countries*. The Nuffield Council on Bioethics, London.

OECD (1986).
http://dbtbiosafety.nic.in/guideline/OACD/Recombinant_DNA_safety_considerations.pdf.

OECD (1992). *Safety Considerations for Biotechnology. Organization for Economic Co-operation and Development*. Paris, France.

OECD (1993). Safety considerations for biotechnology scale-up of crop plants.
http://www.oecd.org/pdf/M00034000/M00034525.pdf.

Rio Declaration on Environment and Development (1992).
http://www.unep.org/Documents.multilingual/Default.asp?DocumentID=78&ArticleID=1163.

UN World Charter for Nature (1982). http://www.un.org/documents/ga/res/37/a37r007.htm

US EPA (1998). Guideline for Ecological Risk Assessment. EPA/630/R-95/002F.

Risk Assessment and Management – Human and Animal Health

ABSTRACT

This chapter discusses the factors that should be considered in making a risk assessment related to the consumption of GM crops by humans as food and by livestock as feed.

OUTLINE

I. INTRODUCTION

As shown in Chapter 1, GM technology is in widespread use in the production of corn, cotton, soybeans and canola, and is likely to be increasingly used in other crops. Other than cotton, these crops are primarily used as animal feed or processed for use in food. For example, yellow corn is used globally as animal feed while white corn is a staple human food in many parts of Africa and Central America. Soybeans and canola, and their derivatives, are important components of many foods. Therefore the potential impact of any GM crop on human and animal health must be assessed during the regulatory process.

The impacts of GM crops on human and animal health are addressed through a risk assessment that considers:

1. the nature of the introduced protein(s) and its potential effects on humans and animals; and
2. the phenotype of the GM plant and whether it has been significantly altered during transformation in ways that could affect human health.

Potential categories of *hazard* can be identified based on what is known about how humans and animals react to food; certain foods are known to be toxic or cause allergies, so these are particular concerns. *Exposure* to the GM crop is assessed based on knowledge of how that crop is used in the food supply. In the risk assessment, GM crops and food derived from these crops are compared with non-GM food stuffs (the *baseline*). Non-GM foods have a long history of safe use and therefore GM foods should be safe for consumption if they are substantially equivalent (see below) to non-GM foods.

In performing a risk assessment for human health, well-characterized model systems are routinely used, including animal and *in vitro* systems. These systems can provide highly conservative estimates of potential risk, which are important because little or no risk can be accepted with respect to human health and most data cannot be gathered directly on humans. In addition, the nature and properties of the GM proteins can be compared to sets of proteins that are known to be safe or unsafe (that is, either toxic or allergenic).

II. THE BASELINE FOR COMPARISON

The process of risk assessment is never intended to demonstrate "zero risk"; the aim is to show that risks are at an acceptable level based on comparisons with a suitable baseline system. In the case of human food and animal feed, the objective is to show that GM food is *as safe as* non-GM food (Box 3.1).

BOX 3.1

HOW SAFE IS FOOD?

No food is 100% safe. We know that the flavour we associate with various foods like hot peppers and onions is produced by chemicals that the plant is using to defend itself against its natural herbivores and these chemicals are toxic to humans if taken in large doses. However, almost any food, including many foods that we eat every day, such as salt or even water, can be toxic if we ingest too much. A cardinal rule in toxicology is "the dose makes the poison". You can eat a dozen carrots at once with no ill effect, but 400 carrots could kill you. Animal studies rarely reveal the possible effects, or safety, of long-term exposure to the kinds of low doses people may experience.

Therefore we cannot expect GM food to be 100% safe – only as safe as non-GM food. In some cases, GM foods may be healthier if the intended changes are designed that way. For example, genes have been introduced into soybeans that result in changes in the fatty acid profile of the resulting soybean oil, leading to lower saturated fat content.

This leads to the concept of substantial equivalence (Box 3.2).

The concept of substantial equivalence is not devoid of limitations, the most important being that there is no consensus working definition. Assessing substantial equivalence requires choosing a suitable baseline of comparison with a well-documented history of use. If components of a given (non-GM) food are known to cause adverse effects, then those effects need to be well characterized. This does not take into account any unintended changes in the composition of GM food. Any such changes may be of toxicological, immunological or nutritional concern and, therefore, the GM food needs to undergo extensive characterization. In view of this, there has been some effort to develop "minimum lists" of food components to underpin substantial equivalence (Schenkelaars, 2002). However, such efforts are constrained by the fact that such lists cannot be universal and should be flexible, changing over time in accordance with changing needs of processors and consumers and with experience. Furthermore, the nature of comparisons may vary from region to region because of specific consumption patterns, for example related to unique foods in particular regions or unusually high consumption of certain foods. Therefore consumption patterns need to be factored into the safety assessment. In addition to these conceptual limitations are others of a practical nature. For example, it may be difficult to apply the concept in cases where nutritional databases are unavailable for a given population.

BOX 3.2

SUBSTANTIAL EQUIVALENCE

The concept of substantial equivalence was first articulated by the Organization for Economic Cooperation and Development (OECD, 1993) and further developed by various organizations including the European Commission (European Commission, 2003). It is analogous to the familiarity concept (see Chapter 2, Section III.C) but focuses on GMOs used as food or feed. According to this, an existing organism used as food/feed with a history of safe use can serve as an appropriate comparator for the assessment of GM food/feed. Application of this concept may result in identification of similarities and potential differences between the GM crop, food/feed and their non-GM counterpart. The new food may be found (a) substantially equivalent to a conventional counterpart; (b) substantially equivalent except for a few clearly defined differences; or (c) not substantially equivalent. Any significant differences would trigger additional tests. It is thought that the outcome of this comparative approach can further structure the safety assessment procedure, which may include additional toxicological and nutritional testing. Application of the substantial equivalence concept is a *starting* point for the safety assessment. It provides assurance that the GM food/feed may be as safe as, or safer than, the traditional counterpart, or that no comparison can be made because of the lack of an appropriate comparator. Analysis of substantial equivalence involves not only a comparison of the chemical composition between the new and the traditional food or feed, but also of the molecular, agronomical and morphological characteristics of the organism in question. Such comparisons should be made with GM and non-GM counterparts grown under the same regimes and environments. When the degree of equivalence is established as substantial, a greater emphasis is placed on the newly introduced trait(s). Where substantial equivalence cannot be shown, this does not necessarily identify a hazard. Where a trait or traits are introduced with the intention of modifying composition significantly and where the degree of equivalence cannot be considered substantial, then the safety assessment of characteristics other than those derived from the introduced trait(s) becomes of greater importance.

The discussion within the Canadian government (http://www.hc-sc.gc.ca/sr-sr/pubs/gmf-agm/substan_table-eng.php) on revising their Guidelines for the Safety Assessment of Novel Foods emphasizes that the concept of substantial equivalence should be taken as a starting point for risk assessment and not as a decision threshold.

TABLE 3.1   Some examples of naturally occurring toxins in crop plants.

Crop plant	Natural toxin
Beans (especially kidney beans)	Lectins
Cassava	Cyanogenic glycosides giving hydrogen cyanide
Fruit seeds (apple, pear and kernel of apricot and peach)	Amygdalin: can be converted to hydrogen cyanide
Parsnip and sweet potato	Furocoumarins and ipomeamarone: produced in response to insect and fungal attack
Potatoes	Solanin, a glycoalkaloid: in shoots and green potatoes
Rhubarb	Oxalic acid
Zucchini (courgette)	Cucurbitacins: defence against insect damage

III. HAZARD IDENTIFICATION AND CHARACTERIZATION

We can identify a number of potential hazards posed by GM food based on known adverse effects of food and these are related to the introduced proteins(s). However, there also may be effects we cannot easily predict that come through unintended changes that occur during the process of introducing the novel proteins, so additional testing is needed to assess whether biologically significant unintended changes have taken place.

A. Potential Toxicity of the Introduced Protein or GM Food

Plants naturally produce various toxins to protect them from pests and diseases and also under various stress conditions. The concentrations of many of these toxins have been reduced by "trial and error" during domestication and selection of crop species, and more recently by targeted breeding. However, even in modern crop varieties there are various toxins that can cause problems if ingested in large quantities or repetitively (Table 3.1). It should be emphasized that most of these are removed in food preparation and cooking.

The two possible toxicity hazards to be considered for GM plants are:

1. Is the introduced protein (or proteins) toxic? This can be assessed by checking for any similarities between the amino acid sequence of the introduced protein and those of known toxins, and testing the novel protein in model systems.
2. Has the process of genetic manipulation activated a natural toxic protein due to a pleiotropic (see the list of commonly used terms in Appendix A) effect? To assess this hazard, standard toxicity tests need to be undertaken.

It should also be noted that people differ in their tolerance to foods which are generally considered to be non-toxic. This food intolerance is an adverse reaction to some sort of food or ingredients that occur every time the food is eaten, especially if larger quantities are consumed. It differs from food allergy (see Section III.B) in that the immune system is not activated and from toxicity which affects anyone eating the food. Food intolerance is usually caused by the person not producing enough of a particular enzyme or chemical needed for digestion of that food. The most common type is intolerance to cow's milk due to the person having low production of lactase which digests lactose, the sugar in milk. Further information can be found on http://www.eatwell.gov.uk/healthissues/foodintolerance/foodintolerancetypes/. Compounds known to result in food intolerance also can be assessed through compositional analysis.

1. Assessing Potential Toxicity

a. Toxicity of the Source of the Novel Protein As part of developing a GM product and later assessing its safety, the origin of the transgenes should be assessed to ensure that they have no history of toxicity to humans or other animals. This history ideally includes some presence in the food supply and therefore a history of safe consumption. For example, insecticidal proteins from the soil bacterium *Bacillus thuringiensis* have been consumed safely by humans for about 50 years through their use in microbial insecticidal sprays. Therefore these proteins are a logical choice for use in GM crops.

b. Bio-Informatic Comparisons with Known Toxins A second starting point for assessing potential toxicity (and allergenicity) of a novel protein is to consider the nature of the sources and to compare its sequence to databases containing known toxins and allergens. A selection of databases is shown in Box 3.3.

Both the software packages for carrying out appropriate sequence comparisons (for example, FASTA at fasta.bioch.virginia.edu), and the protein sequence databases themselves (for example, SWISSPROT at www.expasy.ch/sprot), are publicly available. These bioinformatic comparisons should be performed as part of the process of choosing a gene and protein for use in a new GM crop. Knowledge of the identity and structures of toxic and allergenic proteins is still growing. Therefore updates of the outcomes of the comparisons may be useful after a certain period of time.

c. Acute Toxicity Testing In cases where proteins have been found to be toxic, they are acutely toxic. Therefore toxicity testing typically is carried out using acute tests on species that are highly sensitive surrogates for humans and domesticated animals. In particular, acute toxicity testing

---

BOX 3.3

SELECTION OF DATABASES ON TOXICITY AND ALLERGENICITY

Distributed Structure-Searchable Toxicity (DSSTox) Database Network (http://www.epa.gov/ncct/dsstox/index.html)

DSSTox is a project of EPA's National Center for Computational Toxicology, helping to build a public data foundation for improved structure-activity and predictive toxicology capabilities. The DSSTox website provides a public forum for publishing downloadable, structure-searchable, standardized chemical structure files associated with toxicity data.

Structural Database of Allergenic Proteins (SDAP) (http://fermi. utmb.edu/SDAP/)

SDAP is a web server that integrates a database of allergenic proteins with various computational tools that can assist structural biology studies related to allergens. SDAP is an important tool in the investigation of the cross-reactivity between known allergens, in testing the FAO/WHO allergenicity rules for new proteins, and in predicting the IgE-binding potential of genetically modified food proteins. Using this Internet service through a browser, it is possible to retrieve information related to an allergen from the most common protein sequence and structure databases (SwissProt, PIR, NCBI, PDB), to find sequence and structural neighbours for an allergen, and to search for the presence of an epitope other than the whole collection of allergens.

The Biotechnology Information for Food Safety Database (BIFSD) (http://www.iit.edu/~sgendel/fa.htm)

The BIFSD is a project of the National Center for Food Safety and Technology. It is a portal leading to a variety of databases grouped under a Food Allergy Information page, an Allergen Sequence Database and other resources.

---

with mice (known as an acute mouse gavage) is routinely carried out for GM products because (1) the small body size of mice permits exposure to relatively high concentrations of a novel protein or GM product per unit of body weight, and (2) mice are physiologically similar to humans with respect to susceptibility to toxins. Compounds known to be toxic to humans can be included as controls in these tests. Based on these tests, a level of exposure that results in no effect (No Adverse Effect Level, or NOAEL) can be calculated and compared with the level of exposure to the GM product that could occur through known consumption patterns. The exposure in the toxicity test should be high enough that there is an adequate safety margin between this NOAEL and the expected exposure level. Subpopulations with particularly high exposure and/or high susceptibility to toxins, such as young children, should be considered within this assessment.

B. Potential Allergenicity of the Introduced Protein or GM Product

Allergies are an immune reaction of sensitized individuals to certain compounds that have been ingested (e.g., foods), inhaled (e.g., pollen and house mites) or contacted through the skin (e.g., certain metals). All food allergens are proteins. Individuals have to be sensitized first to become allergic to an allergen; this consists of two phases: (i) sensitization and (ii) allergic reactions in patients that have already been sensitized. During sensitization, the individual's immune system is primed. In cases of food allergy, this occurs at specific patches inside the tissue lining of the intestines where the food allergens come into contact with immune cells. Specific types of white blood cells inside these patches are involved with the priming process that will ultimately lead to the formation of specific antisera against the allergen. The antisera from allergic patients contain a specific class of antibodies that recognize the allergen, the so-called IgE (immunoglobulin E) antibodies. Once sensitized, the individual's antisera will bind to allergens that will reach the bloodstream (after food ingestion) and the complex of allergen molecules bound by antibodies will be recognized by mast cells. These mast cells will subsequently release compounds that are responsible for the symptoms of anaphylaxis, such as histamine (which may cause anaphylactic shock, including blood vessel widening and lowered blood pressure).

There are eight major food proteins that commonly can act as allergens (Table 3.2).

TABLE 3.2 Common food allergens.

Food	Allergen
Milk	Caseins, β-lactoglobin
Egg	Ovomucoid, ovalbumin, ovotransferrin
Fish	Flesh proteins (parvalbumins)
Shellfish	Flesh and shell proteins
Peanuts (and some other legumes)	Storage proteins
Treenuts (e.g., walnut, Brazil nut)	Storage proteins
Soybean	Storage proteins
Wheat (and some other cereals)	Gluten, other storage proteins

As well as these eight, several other foods have been shown to contain allergens. A comprehensive list can be found at: http://www.eatwell.gov.uk/healthissues/foodintolerance/foodintolerancetypes/soyaallergy/

1. Assessing Potential Allergenicity

a. Allergenicity of the Source of the Novel Protein
As in the case of assessing potential toxicity, the source of the novel gene and protein should be assessed to ensure that they have no history of allergenicity.

b. Bio-Informatic Comparisons with Known Allergens
Comparison of the molecular structure of the novel protein with the structures of known allergenic proteins is an important initial step in assessing potential allergenicity particularly because so much is known about common food allergens. There are various searchable databases for allergenicity assessment including www.allergenonline.com and www.allermatch.org as well as those listed in Box 3.3. The bio-informatic comparisons currently focus on the linear sequence of the amino acids of the proteins. The similarity of both small (at least 6–8 amino acids, 100% identical) and large (80 amino acids, at least 35% identical) segments of the linear sequence is considered to be significant. Positive outcomes may indicate the presence of antisera-binding structures within the novel protein that are cross-reactive with the particular allergenic protein sharing the similarity. It should be noted that these comparisons focus on linear sequences, while in reality these sequences will form three-dimensional structures. In fact, it is assumed that binding sites for allergic antisera can be either linear sequence or the three-dimensional structure which brings together amino acids that are separated from each other in the linear sequence. Research efforts are under way to identify the spatial structures of sites in allergic proteins that bind antisera.

c. In vitro Digestibility and Processing Stability
Degradation of the novel protein under simulated conditions of food processing and digestion is an important next step in assessing potential allergenicity because known food allergens are resistant to digestion. Therefore the underlying assumption is that proteins that are not digested in the stomach pass through the digestive system and reach the immune system in the intestinal wall. The immune system may recognize the new protein as "alien" and, eventually, programme itself to develop an allergic reaction as soon as the new protein comes into contact with immune cells. Alternatively, if the new protein resembles another allergen that the body has already become allergic to, allergic reactions can develop upon immediate contact. Either way, stability is regarded as one of the factors that makes it more likely that a protein will be an allergen.

The protein's behaviour under digestive conditions is commonly tested in the laboratory using *in vitro* methods. Digestion in the stomach is simulated *in vitro* by placing the protein in an acid solution (the acidity resembling that of the stomach fluid) and adding pepsin, an enzyme that is produced in the stomach and degrades proteins. Samples are taken after certain periods of time and the integrity of the added protein is checked. If the protein integrity is lost, it is considered to be degraded. Typically, most proteins are degraded within seconds in this model system, while known allergens may be stable for an hour or more. In addition, another *in vitro* method simulates digestion in the intestines using neutral conditions (rather than acidic) and other protein-degrading enzymes that occur in the intestines.

It should be noted that some allergenic proteins are not stable under digestive conditions. This has been observed, for example, for allergenic proteins that are associated with the "oral allergy syndrome". This acute type of allergy occurs when allergens come into contact with tissues in the mouth after consumption, such as in apple allergy.

d. Testing of the Reaction of Antisera from Allergic Patients with the Novel Protein If the initial analyses of sequence of the novel protein and its digestibility are inconclusive, then binding tests with antisera of allergic patients could help to determine whether the tested item will be "recognized" by the immune system of these patients. The tested item may be, for example, a purified novel protein or an extract of whole genetically modified plant seeds. Sera testing can therefore be used to test either the potential allergenicity of newly introduced proteins or any changes in the intrinsic allergenicity of a genetically modified organism, if the non-modified organism is known to be an allergen. The choice of patients for sera testing will depend upon the allergen to which they are sensitive, which should be the same allergen as is suspected to cross-react with the novel protein or the host organism for the genetic modification. Because the allergies against certain allergens are more prevalent than those against other allergens, the availability of patients and antisera directed against specific allergens may vary. The number of antisera will determine the level of certainty to which a tested item can be determined to be allergenic or not. In fact, the "big eight" food allergens (Table 3.2) cause about 90% of the incidents of food allergy, while a vast number of other minor allergens account for the remainder.

e. Clinical Testing (e.g., Skin Prick Test) with Allergic Patients of the Novel Protein or the Whole Genetically Modified Product If a positive reaction were observed in testing antisera, or if suitable antisera were not available, then people with known allergies could be tested directly. Allergenic reactions then may become visible as red wheals around the place of the "skin prick" with the pertinent allergen. As a further step, food challenges may be considered, in which patients will take foods with or without the allergen of interest and their reactions are subsequently being observed. This last type of experiment may be dangerous and this type of test is not recommended because of ethical considerations. In any case, the likely reality is that a GM product or novel protein with allergenic characteristics as demonstrated by sections a–e should not be approved by regulators or advanced by product developers.

f. Animal Testing of the Novel Protein or the Whole Genetically Modified Product   Several species of laboratory animals are currently being tested for their predictive value of allergenicity though none have yet been validated as suitable models. Examples include the Brown Norway rat, the Balb/c mouse, dog and swine. The Brown Norway rat, for example, is a so-called "IgE-hyperresponder", which means that it shows a high tendency towards producing IgE (immunoglobulin E, the type of antibody that is associated with allergy) in its immune sera against allergens with which it comes in contact, for example through its feed.

The *Codex Alimentarius* guidelines for the food safety assessment of foods derived from genetically modified plants or micro-organisms describe, among others, describe how to assess potential allergenicity (*Codex Alimentarius*, 2003). These guidelines recommend the use of a combination of the various methods described above in a "weighted evidence" approach, especially the source of the newly introduced protein, amino acid sequence similarity with allergenic proteins, resistance to simulated stomach digestion, and screening for binding by antisera from patients who are allergic to a specific allergen (sections a–e). In addition, the estimated exposure of the consumer to the newly introduced protein (such as by consumption of foods containing the protein) should be taken into account, which includes, among others, the effect of food processing and consumption data for the relevant food products by the population. Davies (2005) analysed allergenicity of newly expressed proteins as a risk component of GM organisms and suggested a flow chart for assessing their allergenicity (Fig. 3.1)

The case studies described in Box 3.4 illustrate the effectiveness of testing GM products for potential allergenicity before release and show the need for tests to be made on a case-by-case basis.

C. Possible Changes in Nutritional Value and Other Unintended Effects

GM is being used in attempts to enhance the nutritional value of foods for both humans (see golden rice in Box 1.8) and in animals (Box 3.5).

As well as nutrients, food and feed contain anti-nutrients. For example, phytate, a common component of most seeds and cereals, forms a complex with many important minerals such as iron and zinc, making less of the minerals available for absorption from the digestive tract.

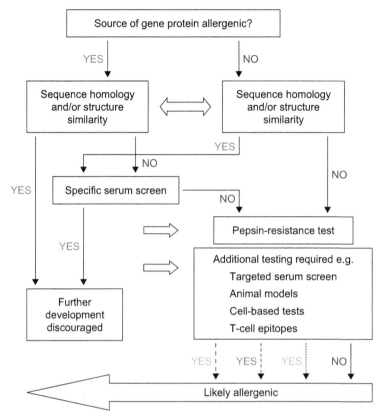

FIGURE 3.1 Flow chart for assessment of allergenicity of a newly expressed protein in a GM organism. Note that the red lettering of "Yes" indicates potential allergenicity and fainter shades of red indicate less potential (from Davies, 2005) (see colour section).

BOX 3.4
_____

CASE STUDIES ON ALLERGENICITY TESTING

1. Soybean and Brazil Nut Protein

In the digestive systems of humans and animals, proteins are degraded to smaller pieces consisting of one or several amino acids that are taken up by the body and utilized. There are certain optimal levels of amino acids that one can use, and therefore certain amino acids may be present in the proteins of foods in limiting amounts. An example of an amino acid that is often limiting in animal feeds is methionine. For this reason, purified

(cont'd)

BOX 3.4   (*cont'd*)

methionine is frequently added to animal feeds to increase the nutritional value of the feed. A few years ago, researchers developed a soybean that was genetically modified with a protein from Brazil nut. This new protein had a particularly high content of methionine and therefore could serve as a substitute for the methionine feed additives. Brazil nut, however, is known as an allergen. In clinical experiments it was observed that patients that were allergic to Brazil nuts also reacted to the GM soybean with the nut protein, but not to conventional soybean. The allergy-causing properties of the Brazil nut apparently had been transferred with the protein to the soybean (Nordlee *et al.*, 1996). The development of this experimental soybean was therefore terminated and the product never reached the market.

2. Pea and Anti-Pest Protein from Bean

Peas (*Pisum sativum*) are susceptible to the pea weevil which can cause considerable damage and loss to the crop. In an attempt to use a GM approach to control this problem the α-amylase inhibitor gene from common bean (*Phaseoulus vulgaris*) was transformed into peas. This gene product would inhibit the digestion of starch by α-amylase in the weevil thereby controlling the pest. However, allergy tests in mice showed that the inserted gene product was allergenic, even though it is not in its natural bean donor. It is thought that subtle changes (different post-translational glycosylation) in the gene product in pea led to it becoming allergenic (Prescott *et al.*, 2005).

BOX 3.5

GM AND ANIMAL FEED

The main foods for animals for meat production are grazing, fodder and food pellets. In intensive meat production pellets based on soybean, corn, sorghum, oats and barley are the main feed. These pellets are supplemented with amino acids, vitamins and minerals for improved nutrition, health and profitability. One of the main limitations on nutrition is that most cereals have a low content of the essential amino acids, lysine and methionine. An anticipated GM product is high lysine corn which should mitigate part of this problem. This should help to reduce the food conversion problem noted in Chapter 1, Section I.A.

Because much of the material for animal feed comes from soybeans and corn, an increasing amount of which are GM (see Chapter 1, Section IV.A) there are issues concerning meat from animals that have been fed GM products. This is discussed in Chapter 5.

Therefore, there is a possibility that the process of transforming a crop plant could inadvertently adversely affect its nutritional value or introduce an anti-nutrient. These unintended changes are not unique to the GM process and may also occur, for example, during conventional breeding of crops. In such cases of unintended effects in crops, breeders try to remove plants showing abnormal phenotype (appearance, behaviour) by selection of new plant breeding lines. In some cases mutant crops with unintended effects have been selected and further developed because of their desirable characteristics.

The nutritional value of a GM product can be established in two ways:

- Possible changes in nutritional value may become evident when comparing data on macronutrients (e.g., fats, proteins and carbohydrates), micronutrients (e.g., vitamins and minerals) and anti-nutrients in the GM plant with those in a comparator non-GM plant. These data are obtained from compositional analyses (see below).
- Additional feeding studies on animal feeds can be conducted with target animal species (such as rapidly growing broiler chickens and lactating dairy cows) to determine factors such as bodyweight increase or milk yield in relation to feed intake (see Case Study 3, Appendix D).

1. Assessing the Composition of the GM Product

Compositional analyses are performed on macro- and micro-nutrients, anti-nutrients, toxins, allergens and compounds from relevant metabolic pathways. As such, these analyses provide an assessment of whether unexpected and unintended changes have occurred with respect to any of these characteristics during the transformation process. These key parameters differ between organisms and information on major crops can be found on various websites. The International Life Sciences Institute (ILSI) hosts an Internet website (www.cropcomposition.org) with compositional data for a number of crops, obtained from analysis of samples from field trials in recent years. The data that are entered in this database have been checked for their quality. The website user can select compositional data with respect to, for example, crop, location, component (nutrients, anti-nutrients, toxins) and year. A further source of information on specific crops is the OECD reports from the Task Force on the Safety of Novel Foods and Feed which can be found on a link from the ILSI website. These reports give much background information on the crops that have so far been considered.

The analysis of specific plant parts may be required, such as whole maize plants for animal feed (because silage is made of them) and kernels for human consumption (starch production, as well as consumption of whole sweet maize kernels). Analytical methods for composition analysis should preferably be standardized methods (such as those of the

Association of Official Analytical Chemists) and done according to Good Laboratory Practice.

Compositional comparisons should be carried out between the GM product and an appropriate counterpart which has a history of safe use in a substantial part of the population. Ideally, to single out effects of the genetic modification from other effects of breeding, the comparator should have the same genetic background to the GM plant. In crops, for example, this should be the near-isogenic line of the genetically modified variety, such as the parental lines. For example, GM insect-resistant maize that has been modified with new insecticidal protein may be compared to conventional non-modified maize that has been obtained from "parents" that were almost the same as that for the genetically modified maize, but without the additional "foreign" genes. Additional reference lines (commonly used commercial lines) should also be included in the comparisons to provide a background range for the variations that may occur in the values for the various measured components. In addition, "historical" ranges for compositional values may be taken from literature for comparison. It should, however, be kept in mind that analytical methods may have changed over time, as well as that crop varieties and their compositions have changed as well. This may render "historical values" less informative. In the case of crops, they should be grown in multiple locations and during multiple seasons to account for the geographic, climatic, environmental, and other factors that may influence outcomes of the compositional assessment.

The outcomes of the compositional analyses should be analysed by statistical methods to detect statistically significant changes in the composition of the genetically modified product versus its comparator(s). However, changes that are found may not necessarily raise safety issues. For example, the changes may be within the background range of values for the specific item that was analysed and thus reflect the levels that consumers are commonly exposed to by this item. In addition, the item itself may not cause adverse effects at the levels that have been analysed. Thus, it is important to identify differences that are relevant to safety.

2. Additional Means of Assessing the GM Product for Unintended Changes

Other important information with regard to unintended effects includes:

- DNA insertion site: Does the inserted DNA interfere with any normal plant gene, especially housekeeping genes and sequences that regulate gene function? If so, is this significant?
- Function and substrate specificity of introduced enzymes: Are other reactions possible besides the intended reactions?

- Phenotypic changes: Does the GM plant differ from its comparator in features such as appearance and development (see discussion in Chapter 4)?

The currently commercialized GM crops contain comparatively simple modifications. In addition, these modifications, in most cases, do not interact with the intrinsic metabolism (biochemistry) of the host and if proteins are formed from new genes, they usually occur at low levels in tissues of the GM crop. By contrast, future generations of GM crops may include crops with more complicated modifications, such as the introduction of new biochemical pathways that lead to the formation of chemical substances that previously did not occur in the particular crop or crop tissue. The potential incidence of unintended effects in these future crops is probably higher than in the current ones.

The question may arise if the comparative assessment as it is currently done would be enough to detect any unintended effects that potentially may occur in future generations of GM crops. Up to now, "targeted" analytical methods that measure specific predetermined substances have been applied to the analysis of genetically modified products. This assessment is far more rigorous than for conventionally bred crops that may have undergone similar unintended changes.

For future crops, profiling techniques may supplement the current targeted techniques for detecting unintended effects. Profiling techniques are holistic analytical methods that may be used to consider whole groups of genes, proteins, or chemical substances. The components that are being analysed need not be known beforehand. Profiles are made of a sample, such as an extract of a GM crop. Such a profile may consist of a number of peaks, for example, in a graph, in which each peak represents a signal generated by a compound that has been separated from others by analytical separations. If this profile of a GM product is compared to that of its comparator, differences in the presence of peaks or the size of peaks may be identified. Peaks of interest may then be selected for further investigation into the identity of the relevant compounds. Examples of profiling techniques are:

- Gene expression micro-arrays for messenger RNAs (genomics); micro-arrays are microscope slides with numerous miniscule spots, each of which contains a probe for a specific gene. Copies of the "messenger RNA", which is generated from active genes which are labelled with a dye, are added to the slide, and the messengers that are derived from the genes for which the probes are specific will bind to these probes. After the rest of the non-bound messenger RNA has been washed away, the bound messenger RNA can be viewed. The location of the specific probe indicates which gene is represented by the bound messenger RNA. This in turn indicates which genes are active in the organism.

- Two-dimensional gel-electrophoresis of proteins (proteomics); protein mixtures, for example extracted from a plant tissue, may be separated in a gel that has been brought into an electric field. Proteins may be separated in the first direction on charge, and after the electric field has been turned 90 degrees to the second direction, on size. After separation, the gel can be taken out and incubated with protein staining dyes to visualize the separated proteins that are present at different sites in the gel. Eventually, the specific sites in the gel may be cut out and the proteins extracted from the gel and analysed further for their identity by another technique, such as mass spectrometry that determines the molecular mass of the particular protein.
- Liquid chromatography coupled to nuclear magnetic resonance for chemical compounds (metabolomics); the "metabolites", chemical substances, in a biological sample can be separated from each other by liquid chromatography by adding the sample on top of a column filled with a solid material with fluid. By subsequently adding more fluid, the metabolites that bind less tightly than others to the solid substance come out more quickly from the column at the outlet. The difference in affinity of the metabolites for the solid compound thus causes them to separate from each other. Different fractions of the fluid coming from the column may be collected and are analysed by nuclear magnetic resonance. A spectrum can then be made of the signals from these fractions within the frequency range of interest.

In all these cases, samples from a GM source would be compared with samples from the corresponding controls and differences identified, such as microscopic spots, protein bands, or nuclear magnetic resonance peaks that are changed in one sample in comparison to the other.

Currently, profiling methods are not operational yet in the safety assessment of GM products. Validation of these methods is needed, as well as data on the background variation, to ensure that the measured data are accurate and comparable, as well as to enable the identification of changes that are outside the natural range. In addition, these techniques generate a large amount of data, which will require filtering to single out those results that are relevant to the safety of foods and feeds. These new approaches would require sophisticated laboratory facilities. If they are to be widely adopted it is likely that they will be undertaken at international centres.

D. Antibiotic Resistance Markers

In Chapter 1, we described the use of selection markers, among which antibiotic resistance genes (Table 1.2) have been widely used. Antibiotic resistance among pathogenic micro-organisms is an important concern and among the main suspected causes of this is the overuse of antibiotics in human therapy as well as the use of low doses of antibiotics in animal feed to promote growth. Transfer of antibiotic resistance genes from GM organisms to pathogenic micro-organisms may, in theory, contribute to antibiotic resistance, though available evidence does not support this view (EFSA, 2004). Most of the DNA in food and feed is degraded during passage through the stomach and intestines. Even so, it is considered to be a risk not worth taking for antibiotics that have significant use on humans and animals, and the use of such antibiotic resistance markers is being phased out. In the meantime it is suggested that antibiotic resistance markers be classified into three groups (EFSA, 2004):

- Group 1: *Unlimited use.* The antibiotic resistance genes in this group are widespread in nature and their corresponding antibiotics are seldom or never used in medicine. The nptII gene (kanamycin resistance) falls under this category.
- Group 2: *Not to be used in commercial GM plants.* Plants possessing these genes may be planted in field trials, but they may not be used for agriculture. These genes confer resistance to antibiotics used in human and veterinary medicine to treat specific infections. These resistance genes are widespread in micro-organisms. It is suggested that growing plants with genes from this group would not increase the distribution of resistance genes in nature. The ampicillin resistance gene fits into this category.
- Group 3: *Not allowed.* These marker genes should no longer be used in GM plants. They confer resistance to antibiotics that are of high value, particularly in human medicine. The effectiveness of these antibiotics must not be compromised. One example of this type of gene is the nptIII gene (resistance to amikacin).

In addition, techniques are now available to remove selection markers from GM lines so marker-free GM crops can be produced.

IV. EXPOSURE ASSESSMENT

As pointed out in Chapter 2, risk is a combination of hazard and exposure. The previous section discussed potential hazards of GM food and feed. Thus, to undertake a risk assessment one needs to understand how and to what extent humans and animals may be exposed to the novel protein(s) and the GM crop. Several factors need to be considered in estimating exposure.

A. Use of the GM Crop

The GM crop can be used directly as food, as processing product for food or as feed for animals; the latter can use different parts of the plant to that used for food; for instance, maize leaves and stalks can be fed directly to animals or may be used to make silage (fermented high-moisture fodder). These uses usually do not differ from those of non-GM crops and therefore information on non-GM crops can be used in assessment estimation. There can be differences in usages (and hence exposure) in subgroups of a population and some subpopulations (e.g., infants) might be particularly sensitive to certain products. There are also differences in digestive systems (ruminants and non-ruminants) of the types of animal which use GM feed; there may be differences in the uptake of the GM protein between the two systems.

B. Amount of Trait Protein in GM Product

The amount of the novel protein that is consumed also depends on the level of expression of that protein in target tissues. This may vary in crops grown in different environmental conditions. Therefore expression levels need to be measured under various environmental conditions (multiple locations) and in all plant tissues that may eventually be consumed in human food or animal feed. Expression levels also need to be measured over the course of the growing season for the relevant crop because protein production and turnover varies with the stage of the plant. Within the risk assessment, the highest observed expression levels in the relevant tissues should be used in calculating margins of safety to ensure that these estimates are protective. Processing the GM product can also alter the amount of the trait protein(s). It could be selectively enhanced or could be reduced to almost zero (e.g., in vegetable oils or in sugar).

V. ASSESSMENT OF STACKED TRAITS

In Chapter 1 we mentioned that GM plants with individual traits (on which food and feed safety assessments had already been made) were being crossed to produce crops with multiple (stacked) traits. Therefore, there is the possibility of interactions between the proteins of the two parental traits and also of unintended effects caused by the interactions of the two parental genomes. A starting point for assessing such products is the assessment of the individual novel proteins. Where no significant risks are found for the individual proteins and events, the stacked product can be compared back to the individual proteins and events to determine whether these single-gene risk assessments will be applicable to the stacked trait product. This comparison requires characterizing the proteins produced in both cases, measuring expression levels of the novel proteins, and characterizing the genetic insertions. If the single-gene and stacked gene products are comparable in these respects, and if the introduced proteins are not expected to interact, then the single-gene assessments can be used and only limited additional characterization of the stacked gene product may be needed. This is more likely to be the case where the stacked gene product is produced through conventional breeding between the single-gene events and where the traits involved work through very different mechanisms (for example, stacking of an insecticidal trait with one designed for herbicidal tolerance). Conversely, if the stacked trait product is produced as a new transformation event, or the traits are expected to interact in complex ways (for example, two traits that involve the same plant metabolic pathway), then plants with stacked traits should be assessed for food and feed safety as though they were new transformed lines. Case studies 1 and 2 (Appendix D) deal with stacked traits.

VI. MANAGEMENT OPTIONS AND MONITORING

The risk assessments of GM food and feed undertaken using information from the potential hazards and exposure described in the sections above help to identify any potential risks. Where potential risks are identified, they can be targeted for management to reduce or preclude them. Risk management will be most effective when it is performed proactively as part of the product design process. Proactive management options can focus on product design and choice of crop/trait combinations. For example, early stage bio-informatic analyses can ensure that the novel proteins being produced in new GM products are not related to toxins and allergens. However, management options are also possible after a product has been approved for commercial use. It is possible to limit products from going into parts of the food or feed supply by channelling, and this has been done both within and among countries. However, management options that truly preclude exposure are preferred because approaches like channelling require rigorous and resource-intensive enforcement and still have not always been 100% effective. Starlink corn (see Box 3.7) is one well-publicized case of attempted channelling that was not successful.

As noted above, the possibility of unintended effects being produced during the creation of a GM crop also should be considered. It is obviously more difficult to monitor for unintended effects and unknown risks than for identified risks. Monitoring human populations occurs in other settings (for example, novel technologies such as mobile phones) where potential risks are well understood. In the case of monitoring for unexpected effects, as would be the case for GM, the value is limited because it is not clear what endpoints and what baseline for comparison should be chosen.

Many countries have organizations that monitor the safety of food produced by conventional means. As there are no differences from conventional food, the potential risks from GM food could be monitored by engaging the same system.

VII. CONTROVERSIES

There have been various controversies about GM products in food which fall under two headings: toxicity of the GM product itself and "contamination" of conventional food with unauthorized GM products. Examples of these controversies are illustrated in the case studies shown in Boxes 3.6 and 3.7.

The examples in Box 3.6 demonstrate the problems associated with segregation of GM and non-GM food supplies.

BOX 3.6

CASE STUDIES ON CLAIMS OF TOXINS IN GM FOOD

1. Insect-Resistant Potatoes

In 1998, Dr. Pusztai announced in the media results of toxicity testing with potatoes which had been modified by the insertion of an agglutinin (lectin) gene (GNA) from snowdrop (*Galanthus nivalis*) to give resistance to the aphid, *Myzus persicae*. This was followed by the publication of these results in a peer-reviewed journal (Ewen and Pusztai, 1999). In the experiments, rats were fed on GM potatoes and on potatoes supplemented with GNA, and histological analyses showed that the GM potatoes had variable effects on the rat gastrointestinal tract which were attributed not only to the GNA but also to the rest of the construct.

The controversy that ensued in the media led to the Royal Society holding an enquiry on the publication (Royal Society, 1999). This enquiry concluded that the experiments were poorly designed, the controls were not adequate, there were too few replicates, the data analysis was improper, there was no account taken of inconsistencies between experiments, and the rats were fed only on potatoes and not the rest of their normal diet.

Thus, the results were considered to be inconclusive.

2. Insect-Resistant Corn

Between 2003 and 2005, a corn event containing the Bt Cry3Bb1 toxin for protection against corn rootworm was authorized for human consumption in a number of countries including Australia, Canada, China, the EU, Japan, New Zealand and the USA. In each of these countries, the competent authorities assessed the feed studies data supplied by the developer. Following a court case in Germany, one particular study – a 90-day rat-feeding study – was placed in the public domain. From an independent

(cont'd)

BOX 3.6 *(cont'd)*

analysis of these data (Séralini *et al.*, 2007), it was claimed that there was evidence of liver and kidney toxicity in the rats and therefore that the safety of this GM corn for human consumption was questionable. This study was released to the press who publicized it extensively. Several national and regional food safety authorities (e.g., The European Food Safety Authority and The Food Standards Australia New Zealand) re-examined their original assessment and concluded that their original opinion was valid. They questioned the statistical analyses used in the independent study and the lack of any new data.

3. Herbicide-Tolerant Soybeans

Reports of feeding studies indicating that a glyphosate-tolerant soybean line compromised the fertility of rats and the survival of their offspring led to further controversy. The experiments were never published in a peer-reviewed journal but some details were divulged in interviews published in *Nature Biotechnology* (Marshall, 2007). In this article and in subsequent letters to that journal, various features of the experimental approach were criticized.

All three of these case studies show that studies on possible deleterious effects of GM products should be carefully peer reviewed before being put into the public domain.

BOX 3.7

CASE STUDIES ON "CONTAMINATION" OF GM FOOD

1. "StarLink" Corn

"StarLink" was a GM corn variety that contained the Bt protein Cry9C to confer resistance to the European corn borer. It was given conditional registration by the US Environmental Protection Agency (EPA), with the registration limiting its use to animal feed. It was not approved for human use because the Cry9C protein was considered to be a potential allergen. It was also not allowed to be shipped overseas.

(cont'd)

BOX 3.7   (*cont'd*)

In 2000, StarLink was grown on 352,000 acres of land and represented 0.5% of the 10.4 billion bushel US corn crop.

However, the GM corn was detected in taco shells distributed by three companies and turned up in various countries including Japan who prohibited GM corn for both food and feed. Thus there was a clear violation of the conditions of the EPA registration – to keep the GM corn out of the food supply and ensure there was no export.

Additionally, the Cry9C protein turned up in a corn hybrid produced in 1998. This was suggested to be due either to cross-pollination in the field or to mishandling of the seed during production and distribution.

The damage was not limited to potential risk to human health and the environment but also extended to the economic costs of the recall of hundreds of products, the removal of the contaminated food products from food outlets, and the destruction of corn crops on hundreds of thousands of acres as well as the potential loss of valuable export markets. There were also potential liability issues (see Chapter 5).

2. LLRICE 601

LLRICE 601 is a long-grain rice genetically modified with the phosphinothricin-N-acetyltransferase gene to confer tolerance to the herbicide glyphosinate ammonium herbicides. It was not intended for commercial use and was given permission for experimental field testing between 1998 and 2001 with the restriction that it was not for human consumption.

In 2006, traces of LLRICE 601 were found in commercial rice in the USA. The following shows how rapidly the controversy about this spread worldwide and various reactions to it:

18 August 2006: US Agriculture Secretary Mike Johanns announced that US commercial supplies of long-grain rice had become inadvertently contaminated with a GM variety not approved for human consumption. (Note that this announcement was made late on a Friday evening.)

19 August: Japan suspends imports of US long-grain rice following a positive test for trace amounts of a GM variety not approved for human consumption.

21 August: EU demands more details of tainted US biotech rice.

21 August: Greenpeace demands global ban on US rice imports.

23 August 2006: EU says rice from US needs proof not contaminated.

24 August: EU adopts tough rules for US long-grain rice.

30 August: American farmers sue company over GM rice.

7 September: USDA moves to deregulate controversial GM rice.

References

Codex Alimentarius (2003). Codex principles and guidelines on foods derived from biotechnology. Rome: Codex Alimentarius Commission Joint FAO/WHO Food Standards Programme, Food and Agriculture Organization.

Davies, H.V. (2005). GM organisms and the EU regulatory environment: allergenicity as a risk component. *Proc. Nutrit. Soc.* **64**, 481–486.

EFSA (2004). http://www.gmo-compass.org/pdf/documents/efsa_marker.pdf. See also GMO Compass for a classification of marker genes. http://www.gmo-compass.org/eng/safety/human_health/126.position_efsa_antibiotic_resistance_markers.html

European Commission (2003). Communication from the Commission on the Precautionary Principle. COM (2000) 1 final.

Ewen, S.W.B. and Pusztai, A. (1999). Effect of diets containing genetically modified potatoes expressing Galanthus nivalis lectin on rat small intestine. *Lancet* **354**, 1353–1354.

Marshall, A. (2007). GM soybeans and health safety – a controversy re-examined. *Nature Biotechnol.* **25**, 981–987.

Nordlee, J.A., Taylor, S.L., Townsend, J.A., Thomas, L.A. and Bush, R.K. (1996). Identification of a Brazil-nut allergen in transgenic soybeans. *New Eng. J. Med.* **334**, 688–692.

OECD (1993). Safety evaluation of foods derived by modern biotechnology: concepts and principles. http://www.agbios.com/docroot/articles/oecd_fsafety_1993.pdf

Prescott, V.E., Campbell, P.C., Moore, A., Mattes, J., Rothenberg, M.E., Foster, P.S., Higgins, T.J.V. and Hogan, S.P. (2005). Transgenic expression of bean α-amylase inhibitor in peas results in altered structure and immunogenicity. *J. Agric. Food Chem.* **53**, 9023–9030.

Royal Society (1999). Review of data on possible toxicity of GM potatoes. http://royalsociety.org/document.asp?tip=1&id=1462

Schenkelaars, P. (2002). Rethinking substantial equivalence. *Nature Biotechnol.* **20**, 119.

Séralini, G.E., Cellier, D. and de Vendomois, J.S. (2007). New analysis of a rat feeding study with genetically modified maize reveals signs of hepatorenal toxicity. *Arch. Environ. Contam. Toxicol.* **52**, 596–602.

Further Reading

Dona, A. and Arvanitoyannis, I. (2009). Health risks of genetically modified foods. *Crit. Dev. Food Sci. Nutrit.* **4**, 164–175.

Guidance Document for the Risk Assessment of Genetically Modified Plants and Derived Food and Feed (2003). European Commission, Health and Consumer Protection Directorate-General, The Joint Working Group on Novel Foods and GMOs. http://ec.europa.eu/food/fs/sc/ssc/out327_en.pdf.

Hammond, B.G. (2008). *Food Safety of Proteins in Agricultural Biotechnology*. CRC Press, Boca Raton, FL.

Kuiper, H.A., Kleter, G.A., Noteborn, H.P.J.M. and Kok, E.J. (2001). Assessment of the food safety issues related to genetically modified foods. *Plant J.* **27**, 503–528.

Kuiper, H.A., Kok, E.J. and Engel, K.H. (2003). Exploitation of molecular profiling techniques for GM food safety assessment. *Curr. Opin. Biotechnol.* **14**, 238–243.

Risk Assessment and Management – Environment

ABSTRACT

This chapter describes the key elements of assessing the risk to the environment of releasing a GM product. Potential hazards include impacts of the introduced gene on non-target organisms and adverse effects of gene flow to sexually compatible species. Various approaches to risk management and monitoring are discussed.

OUTLINE

I. INTRODUCTION

A. The Basics of Environmental Safety Assessment

As shown in Figure 1.8, GM crops such as corn, cotton, soybeans and canola are grown on large areas. Consequently, GM crops have the potential for broad environmental impacts both because of deliberate changes made to the crop and indirect effects of these technologies on other agricultural practices. The potential impact of any GM crop on the environment therefore must be assessed prior to its commercialization.

There are two elements of the environment that agricultural technologies like GM crops can impact: the agricultural environment and the non-agricultural environment. The agricultural environment is that cultured by humans and is typically made up of relatively few plant species often grown as monocultures or intercropping with a limited number of species; as noted below, it is influenced considerably by agricultural practices. The non-agricultural environment is the so-called "natural" ecosystem, comprising a variety of intermingled plant and animal species, but it should be recognized that humans have to a greater or lesser extent also influenced this environment.

The overall approach to environmental safety assessment is shown in Fig. 4.1.

There are two overall focal points for the assessment: the GM plant (the product) and the introduced gene/protein (the trait).

As various specific terms are used in environmental risk assessment, definitions are given in Box 4.1.

B. Specifics of how Risk Assessment Applies to Potential Environmental Impacts

In Chapter 2, we discussed that the process of risk assessment comprises five interlinking stages: hazard identification, exposure assessment, risk characterization, risk management and monitoring. The overall approach for environmental assessment is to:

- Consider the nature of the introduced protein(s) and separately assess the phenotype of the GM plant;
- Identify a baseline for GM food and feed; this is usually non-GM crops grown under normal commercial conditions;
- Identify potential hazards based on what is known about how the crop and the novel protein interact with the environment, most obviously through possible toxicity to wildlife;
- Assess exposure using knowledge of the crop and how it is grown;

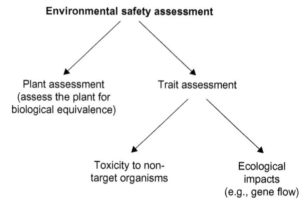

FIGURE 4.1   Overall approach to environmental safety assessment.

BOX 4.1

DEFINITIONS

Ecology: The study of the interrelationships between organisms and their environment

Ecosystems: A community of living organisms and the environment in which they live, interacting to form a whole functional system

Food web: A complex intermeshing of individual food chains in an ecosystem. A food chain is a sequence of organisms, each of which uses the next, lower member of the sequence as a food source

Gene flow: The movement of genes (i.e., alleles) within a population or between interbreeding populations as a result of outcrossing and natural selection or seed migration

Non-target organisms: Any organism for which pesticidal control was either not intended or not legally permitted by application of a pesticide or other product (e.g., industrial chemical) that affects that organism

Tiered testing: A system in which tests are performed at one level and those that do not pass at that level are further and more stringently tested at higher levels

---

- Risk management can be performed proactively, and environmental monitoring is routinely carried out after commercialization of a GM crop.

These steps are dealt with in more detail in the sections below.

C. Model Systems to Represent Worst Case

For an understanding of potential risk of any new trait and how to deal with it, the worst case scenario has to be considered. As this has to be prospective (before the event) rather than retrospective (after the event), model systems are often used. Such model systems for understanding hazards to the environment typically use well-characterized indicator species and high levels of exposure, in the same way that model systems are used in assessing the risk of a GM food being toxic or allergenic to humans (see Chapter 3).

As discussed in Section III.A.1, threatened and endangered species that may be impacted by the introduction of a GM crop require special consideration, as do crop species in their centres of origin and diversity.

II. THE BASELINE FOR COMPARISON

We pointed out in Chapters 1, 2 and 3 that risk assessment of GM crops should be made against a baseline which is generally accepted to be the non-GM version of that crop grown in that region. All agriculture has complex impacts on the environment (Table 4.1) which necessitate the use of an appropriate baseline.

A starting point for obtaining a baseline is "familiarity" (see Chapter 2, Section II.C), which is rooted in our knowledge and experiences of agriculture and particular crops in that region. This knowledge helps us to

TABLE 4.1   Some impacts of conventional (non-GM) agriculture on the environment.

Factor	Agricultural environment	Natural environment	Overall environment
Agricultural practice	Enhance yield Loss of fertility Land degradation Reduction in sustainability	Loss of natural environment Impact on landscape Loss of biodiversity	Reduction in carbon fixation Use of fossil fuels in fertilizer production
Inputs (pesticides and herbicides)	Reduce biotic losses Resistance of target Loss of biodiversity	Little apart from local spread	Contamination of human and animal food Use of fossil fuels in production
Inputs (fossil fuels)	Cost	Little impact	Carbon emissions
Water	Salinization Drought	Water depletion Water pollution	Water depletion Water pollution

understand typical patterns of variability in common biological characteristics of the crop plant being considered. In most regions, non-GM crops have been grown for a considerable time and so we can determine whether the interactions between these crops and the environment, including the typical growth patterns and ecological interactions, are likely to be changed in GM crops. To this end, the OECD has produced consensus documents on the biology of common crops (see www.oecd.org). This reliance on experience and history of safe use is comparable to the use of substantial equivalence in assessing potential health risks associated with GM foods, and therefore familiarity can also be thought of as a sort of "ecological equivalence".

In many cases, the introduction of a GM trait also leads to the agronomic system being adapted to take advantage of the trait. An example of this is the increased adoption of no-till or low tillage agriculture by farmers using herbicide tolerant crops (see Box 4.9). Thus, GM cropping systems need to be compared with conventional agricultural systems as practised in the region.

Agriculture of any type has some impact on the environment; biodiversity typically is relatively low in agro-ecosystems precisely because

---

BOX 4.2

COMPARISON OF PEST CONTROL BY CONVENTIONAL INSECTICIDES AND GM TECHNOLOGY

A recent study by Wolfenbarger *et al.* (2008) compared the abundance of various non-target arthropods (predators, parasitoids, omnivores, detritivores and herbivores) on Bt and non-Bt cotton, corn and potato that had or had not received insecticide treatments. The data were a compilation of published literature (and can be found in a public database) and covered various types of Bt toxins. Overall, insecticidal effects were much greater than the effects of the Bt crops. In addition, insecticide use was higher on the non-Bt crops and therefore the insecticide impacts on non-Bt crops were larger than those on Bt crops, but the type of insecticide also influenced the magnitude of effects. There were specific effects of the Bt crops when compared with the unsprayed non-Bt crop, but mostly these were expected reductions in the populations of specialist parasitoids of the target pest. In any case, pest problems in non-GM crops are usually addressed by spraying insecticides and so unsprayed non-Bt crops do not necessarily represent the real situation. The authors suggest that "these results will provide researchers with safety information to design more robust experiments and will inform the decisions of diverse stakeholders regarding the safety of transgenic insecticidal crops".

these systems are managed to maximize crop yields. Therefore GM crops must be compared with non-GM counterparts to understand whether adverse effects may occur, including through changes in the agronomic practices that typically occur in that crop. For example, GM crops with insecticidal proteins need to be compared to non-GM crops with alternative pest controls in place such as insecticidal sprays (Box 4.2).

III. HAZARD IDENTIFICATION AND CHARACTERIZATION

We can identify a number of potential hazards posed by GM crops based on known properties of the non-GM crop and the introduced protein(s). There also may be effects we cannot easily predict that come through unintended changes which occur during the process of introducing the novel proteins; so additional testing is needed to assess whether biologically significant unintended changes have taken place.

Potential hazards to the environment can be divided into:

- Impacts of the introduced protein;
- Effects of other unintended phenotypic changes (RNA silencing may also lead to these); and
- Indirect impacts on the environment related to the introduction of the GM crop.

A. Potential Adverse Impacts of the Introduced Protein

1. Potential Toxic Effects of the Introduced Protein on Wildlife (Non-Target Organisms)

The novel protein in the crop plant may be toxic to organisms that feed on it either deliberately (such as herbivores) or inadvertently (as with predatory organisms feeding on herbivores on a GM plant). Crops expressing pesticidal proteins that target specific agricultural pests constitute one of the largest categories of GM crops developed to date. A particular concern about GM crops containing the insecticidal Bt proteins (Box 4.3) is that they might affect organisms other than the target pest (non-target organisms (NTOs)).

Affected non-target organisms potentially could include beneficial insect species and soil organisms, non-target pests, and species that are deemed to be important for the maintenance of biodiversity. The issue here is whether the expressed pesticidal toxin may significantly reduce populations of beneficial species. However, the same concern also applies in the case of chemical pesticides used for the same agronomic purpose.

Indirect effects on non-target organisms also need to be considered. For example, the removal of a target pest may deprive beneficial

BOX 4.3

BACILLUS THURINGIENSIS (BT)

Bacillus thuringiensis is a Gram-positive, soil-dwelling bacterium. *B. thuringiensis* also occurs naturally on the dark surface of plants. *B. thuringiensis* is closely related to *B. cereus*, a soil bacterium, and *B. anthracis*, the cause of anthrax: the three organisms differ mainly in their plasmids.

Upon sporulation, *B. thuringiensis* forms proteinaceous insecticidal δ-endotoxins either in crystals (Cry toxins) or cytoplasmically (Cyt toxins), which are encoded by *cry* or *cyt* genes, respectively. When insects ingest toxin crystals, the enzymes in their digestive tract cause the toxin to become activated. The toxin binds to and into the insect's gut membranes forming a pore that results in swelling, cell lysis and eventually killing the insect. *B. thuringiensis* also produces insecticidal proteins at other stages in its lifecycle, specifically the vegetative insecticidal proteins (VIPs).

Cry toxins have specific activities against species of the orders Lepidoptera (moths and butterflies), Diptera (flies and mosquitoes) and Coleoptera (beetles) (see Table 1). Cyt proteins are active against Diptera and Coleoptera.

TABLE 1   Some Cry endotoxins and their specific activities.

		Major target insects
Cry protein	Order[a]	Common name
Cry1Ab, 1Ac	L	Cotton bollworm, tobacco budworm, European corn borer
Cry1F	L	Tobacco budworm, European corn borer, fall armyworm
Cry2Aa, 2Ab	L & D	Cotton bollworm, tobacco budworm, European corn borer, fall armyworm, mosquito
Cry3A, 3B	C	Colorado potato beetle, corn rootworm
Cry4A	D	Mosquito

[a]Insect order: C = Coleoptera; D = Diptera; L = Lepidoptera

Cry and Cyt proteins are not toxic to a wide range of other insects, earthworms, birds and mammals.

Thus, *B. thuringiensis* serves as an important reservoir of Cry toxins and *cry* genes for production of biological insecticides and insect-resistant genetically modified crops. Because of their insecticidal activity, Cry proteins have been used to control insect pests in agricultural systems for almost 50 years, initially as formulated microbial sprays and more recently incorporated into GM crops. Similarly, Cyt proteins have been used in

(cont'd)

BOX 4.3 (*cont'd*)

sprays for mosquito control. The transgenic expression of a Bt Cry gene means that the toxin is produced throughout the crop plant and is very effective against pests that live within the stem (e.g., corn borer) or in cotton bolls (e.g., cotton bollworms) which are difficult to control by conventional insecticides.

Information on the nomenclature of *B. thuringiensis* toxin proteins can be found at www.lifesci.sussex.ac.uk/home/Neil_Crickmore/Bt/intro.html

---

organisms of their food source resulting in starvation and concomitant reductions in their populations. Similarly, indirect effects may occur in the case of certain herbicide-tolerant transgenic crops where the elimination of weeds may deprive herbivorous animals of their food source. Theoretically, these food web effects could extend to birds, reptiles, amphibians and mammals. However, it must be remembered that alternative chemical, biological or cultural methods of pest or weed control are likely to produce similar impacts.

Consideration of effects on non-target organisms can be particularly complex because it is very difficult to establish relationships between the source(s) of risk (cause) and potential environmental harm (effect). Therefore, assessing hazards to non-target organisms involves a *tier-based* system that begins with simple laboratory toxicity tests and proceeds to more complex field tests as needed (EPA, 2007; Romeis *et al.*, 2008).

Laboratory-based toxicity testing involves a range of *indicator* or surrogate species that represent groups of organisms that could be exposed to the GM crop. These surrogate species should be drawn from various taxonomic and ecological groups, including organisms that are closely related to the targeted pests (in the case of insecticidal traits). This use of model systems in toxicity testing is done for some of the same reasons that model systems are used in human health assessments (Chapter 3); the species tested are representative of the groups being assessed, and the test systems are well-worked out to produce reliable results. Therefore, the tiered approach to testing, and the range of organisms tested, will be generally applicable to any GM crop and environment. The actual species tested should consider the specific crop and trait, and the region in which the releases are to occur, but appropriate surrogate organisms may not come from that region itself. As an example of hazard testing for effects on non-target organisms, Box 4.4 shows the standard set of tests required by the US Environmental Protection Agency for insecticidal crops (termed Plant-Incorporated Protectants by the US EPA).

Laboratory testing should be carried out at concentrations of the expressed protein that match or exceed those to which these species

---

BOX 4.4

EPA PRIMARY TESTING

Data are required on the ecological effects on representative non-target terrestrial and aquatic species. The testing goes through four tiers; if a problem is identified on one tier, the testing goes on to the next higher tier:

Tier I: tests typically use purified proteins in an artificial diet and are subchronic:

- Avian oral toxicity test on an upland game bird;
- Freshwater fish oral toxicity test;
- Freshwater invertebrate test on *Daphnia* or aquatic insect species;
- Honey bee test for larval and adult bee toxicity as a representative pollinator;
- Insect predators and parasites, for example green lacewing larvae, the ladybird beetle, and a parasitic wasp;
- Non-target soil insect and/or other invertebrate, typically including *Collembola* and an earthworm species.

If the Tier 1 tests are not possible or show adverse non-target species effects at field use rates, then testing of additional species and/or testing at a higher tier level is required.

Tier II: Testing with plant tissues alone or in an artificial diet

Tier III: Chronic, reproduction, lifecycle and population effects testing:

- Chronic broiler study.

Tier IV: Simulated or actual field testing to determine if there is a noticeable change in wildlife populations under field use conditions

---

would be exposed under field conditions, so that safety margins can be calculated relative to the expected environmental exposure.

Because the laboratory tests are carried out at unrealistically high levels of exposure, an adverse effect under these conditions does not necessarily mean that a risk exists. Work on the impact of Bt corn pollen on the Monarch butterfly illustrates the difficulty of extrapolating from laboratory testing to what happens under natural conditions (see Box 4.5).

Where laboratory tests suggest a potential hazard or where appropriate laboratory tests cannot be performed, additional testing under more realistic exposure conditions is needed. This testing may use GM plant tissues in the laboratory or field testing under a controlled field release (see Chapter 2, Section III.B.3). As in the case of the laboratory testing, field testing should still focus on a number of representative groups of non-target organisms rather than trying to measure impacts on everything.

BOX 4.5

THE MONARCH BUTTERFLY AND BT

In 1998, John Losey and his colleagues published a short letter in *Nature* showing that Monarch butterfly caterpillars had lower survival and slower development when feeding on milkweed leaves coated with pollen from Bt corn plants than when feeding on milkweed leaves alone or leaves with pollen from non-GM corn (Losey *et al.*, 1999). The media took this as a demonstration that the use of Bt corn pollen was decimating monarch populations in the USA.

The GM non-target conclusions were:

- Monarchs are related to the target species (moth pests) so a potential hazard is to be expected;
- Exposure issues are critical: are Monarchs present in corn fields during pollen shed and, if so, are they exposed to toxic levels of the pollen?
- Less than 1% of the population is present within Bt corn fields during pollen shed;
- Absolute and relative risk is very low.

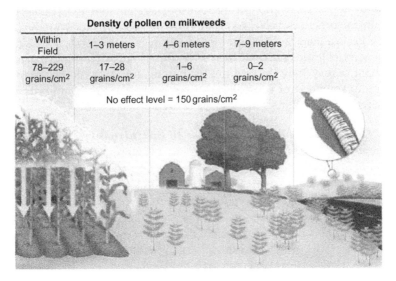

FIGURE 1   Density of pollen on milkweeds. (Preliminary data reported at the Monarch Butterfly Research Symposium by Dr. Mark Sears, University of Guelph, Dr. Galen Dively, University of Maryland and Dr. Rich Hellmich, USDA) (see colour section).

(cont'd)

BOX 4.5   (cont'd)

In reality, the original study only demonstrated a hazard from the presence of high levels of Bt corn pollen, and this hazard was to be expected because the Cry proteins expressed in Bt corn varieties were chosen for their ability to control pest Lepidoptera (moths). Subsequent studies looked at whether Monarch butterflies actually are exposed to sufficient levels of Bt corn pollen in the field to cause adverse effects (Sears *et al.*, 2001).

These studies assessed the distribution of milkweed plants (see Fig. 1) around corn fields, the distribution of corn pollen over space and time, and the distribution of Monarch butterflies around agricultural fields. Based on this evidence, the major varieties of Bt corn that are in commercial use were found not to pose any significant risk to Monarch butterflies. This study illustrates the need to understand field exposure for assessing risk, and the common confusion between hazard and risk.

As with other aspects of the risk assessment, the impacts of the introduced protein should be compared with potential impacts of alternative agricultural practices and on agricultural sustainability. Because most agricultural practices have some adverse effects on biodiversity (agriculture is designed to maximize production of one or a few plant and animal species), the choice of the baseline for comparison is particularly important here. Where the GM trait is insecticidal, it should be compared with other methods of controlling insect pests that are commonly used such as conventional insecticides (see Box 4.2).

2. Potential Selection for Resistance to the Introduced Protein in Targeted Pest Species

Experience with conventional pesticides and herbicides has demonstrated that pests and weeds routinely have the potential to evolve resistance to control strategies, whether they rely on host plant resistance, chemicals, or cultural practices. Thus, where the transgenic protein is targeted at pests or weeds, it is possible that these target species may evolve resistance to that protein. While this is not itself a hazard, if resistance were to evolve, other less favourable and effective technologies would need to be used to control these same pests and weeds, which could adversely affect the environment.

Consequently, the approach that has been adopted for GM crops with insecticidal proteins in most countries is to acknowledge that the possibility exists of pests evolving resistance and to manage it proactively. These risk management programs are described in detail in Section V.

3. Potential Effects of the Introduced Protein on the Fitness of Related Plant Species

The transgene introduced into the GM plant potentially may move into non-GM plants of the same species or into related reproductively compatible species. This process is known as *gene flow* (Box 4.6).

BOX 4.6

GENE FLOW

Gene flow is a natural part of the biology of plant species and is important in the maintenance of genetic variation in populations, as well as in the spread of new traits among populations and across species boundaries. It typically involves movement of pollen and is dependent upon wind or animal vectors (pollinators). Gene flow occurs with all species, and thus with all crop species, but the amount of gene flow is a function of crop biology. The amount of gene flow will depend on how far pollen disperses from the GM plants and whether suitable plant relatives are present within range of pollen movement (also known as pollen flow). Plant species that have pollen that is readily dispersed considerable distances by wind (such as maize) or by insect pollinators (such as canola) will have the greatest amount of gene flow. Likewise, gene flow will be greatest where GM crop plants are grown together with abundant wild or feral relatives, particularly in centres of origin (see Box 1.1).

For gene flow to occur from any particular GM crop field, several conditions must hold:

- Sexually compatible plants of the same or related species must exist near the GM crop;
- These plants must be within the range of pollen dispersal for the GM crop;
- Pollen shed for the GM crop must coincide with receptivity for the recipient plants;
- The resulting offspring must be viable and fertile.

Overall, the potential impacts of gene flow from GM crops are assessed in two steps:

1. The potential for gene flow to occur (likelihood) between the GM crop and any wild relatives is estimated (the exposure component); and
2. The potential environmental impact of gene flow (the hazard component), if it were to occur, is assessed.

Considerable information is already available on the biology of all major crops, making it relatively straightforward to characterize the likelihood

(cont'd)

BOX 4.6   (*cont'd*)

of gene flow for any given crop using published literature and simple field surveys. The rate of gene flow is well known for many species and depends on various factors:

- The degree to which a crop is self-fertilized versus cross-pollinated;
- The number and mobility of propagules (like pollen) produced by each crop plant;
- The presence and abundance of suitable wild relatives.

Gene flow will be higher from crops possessing characteristics that include high pollen production, an ability to disperse pollen over long distances, pollen production over a long period of time, and/or abundant outcrossing wild relatives.

For example, if we consider Europe or North America, crops like oilseed rape (see Table 1) that display many of these characteristics will have relatively high levels of gene flow, and others like cotton, potato and wheat will have low levels of gene flow.

TABLE 1   Gene flow frequency (exposure): examples for Europe (Data from Eastham and Sweet (2002)).

Crops	Crop to crop	Crop to wild relative
Oilseed rape (canola)	High	High
Sugar beet	Medium to high	Medium to high
Maize (corn)	Medium to high	No known wild relatives
Potatoes	Low	Low
Wheat	Low	Low
Barley	Low	Low
Fruits (apple, plum, grape, blackberry, blackcurrant, raspberry, strawberry)	Medium to high	Medium to high

Conversely, many factors may preclude gene flow between GM crops and wild relatives. For example, in reviewing the potential for gene flow from Bt corn, Bt cotton or Bt potatoes to wild relatives in the USA, the EPA concluded that pollination of any wild relatives that might exist would not occur because of a combination of differences in chromosome number, phenology and habitat. This sort of assessment permits attention to be focused on GM crops with a higher likelihood of gene flow (Eastham and Sweet, 2002).

However, gene flow does not, in and of itself, represent a risk because it is only a source of environmental exposure to the transgenic protein; whether there is an adverse impact (hazard) of the transgenic protein in the related plant species must be understood. Possible hazards could include:

1. Adverse effects on non-target species feeding on the relative of the GM crop if the protein is insecticidal; or
2. Changes in the fitness of the relative of the GM crop that affect its abundance in natural plant communities (for example, because it is more resistant to key pests) or in agricultural systems (for example, because it becomes more of a weed through increased competitive ability or tolerance to common herbicides).

Evaluating the potential for adverse effects through gene flow can be done by assessing the level of gene flow (exposure) or by assessing the consequences of gene flow (the hazard). In most cases, determining that interspecific gene flow is absent or negligible is the simplest approach because this can be done by showing that suitable reproductively compatible relatives are not present where the GM crop is grown. Many crops are now primarily grown outside of the region in which they originated so this is often the case; however, as noted later, special consideration should be given to this issue for releases of crops in their centres of origin.

Cases where gene flow through pollen-mediated hybridization could give rise to potential hazards include:

- *Crop-to-crop hybridization.* Spontaneous hybridization leading to the evolution of increased weediness has been shown in the case of numerous non-GM crops (Ellstrand, 2003) and, theoretically, analogous situations could be possible with GM crop plants. In such cases, potential hazards may arise from the stacking of several transgenes as has been observed in the case of the stacking of three herbicide resistance genes from GM-tolerant canola varieties into a non-GM one (Hall *et al.*, 2000). Contamination of the food chain could be another potential hazard if the transgene(s) were coding for compounds intended for industrial or pharmaceutical use.

- *Crop to related weed.* In such cases, evolution of increased weediness would be a matter of concern if the escaped transgene(s) conferred a fitness advantage to the weed. However, this would require the stable incorporation of the novel gene into the chromosome of the weedy plant through repeated backcrossing of the crop/weed hybrids to the new species. Although crop/weed hybrids have reduced fertility, introgression of non-GM crop genes into a new weedy species has been shown to be possible (Snow *et al.*, 2001) and analogous situations could arise in the case of transgenes (see Section III.B.1 for weediness).

- *Crop to the same native species.* Gene flow from GM plant to non-GM plants of the same species generally does not represent an environmental hazard but may be a concern for other reasons. In particular, gene flow between GM crops and local landraces (see the list of commonly used terms in Appendix A) has been identified as a concern in countries where those landraces have special cultural significance such as with corn in Mexico (Box 4.7).

BOX 4.7

BT CORN IN MEXICO

An article published in 2001 reported that traits from GM corn had introgressed into local Mexican landraces of corn, which raised questions about potential long-term effects on genetic diversity and biodiversity (Quist and Chapela, 2001).

The points arising from this paper concerning gene flow were:

- Evidence of GM traits has been found in Mexican landraces of corn, even though these products are not approved for commercial use in Mexico;
- Potential effects on genetic diversity;
- Concerns about how the gene flow occurred;
- Introgression probably has occurred (possibly through deliberate breeding by farmers) but the risks posed are minimal;
- Traits must be assessed case by case.

The article fuelled debate in the scientific community over whether such introgression had occurred and, if so, what its impact would be. Uncertainty remains on the first point because of methodological problems in the original article (see, for example, Christou, 2001), but it seems likely that introgression has occurred though probably at lower levels than first reported. Regardless, scientists have concluded that this is associated with minimal risk to the Mexican landraces or wild relatives of corn like teosinte because the traits involved will have little effect on fitness and introgression with conventional corn varieties has been going on (and encouraged by Mexican smallholders) for a very long time (see Wisniewski *et al.*, 2002). This issue highlights some of the common misunderstandings about gene flow, and illustrates the special technical and cultural concerns related to gene flow in centres of origin.

- It should be remembered that gene flow is a natural process and is going on continually within non-GM crops including landraces. Therefore, complete isolation and preservation of landraces may not be a realistic goal. Furthermore, gene flow is a two-way process between GM and non-GM crops.
- In addition, farmers growing non-GM crops in some countries may market their crop specifically as non-GM, requiring that they minimize gene flow between their crop and GM fields. Whether the responsibility for isolation rests with the farmer growing the GM crop or with the farmers growing the non-GM crop can be debated, and regulations vary among countries. This is discussed in more detail in Chapter 5.

- *Crop to related native species.* Pollen-mediated dispersal of alleles from common crop cultivars and introgression into locally rare landraces or wild relatives might cause the extinction of the latter if the rare population produces a very large number of hybrids and becomes absorbed into the commercial cultivar (known as genetic assimilation or swamping) or if the resulting hybrids have reduced fitness (known as outbreeding depression). Genetic assimilation and outbreeding depression are known to occur in animals but have received little attention in the case of plants. Extinction by hybridization is now recognized to be a conservation problem and has been implicated in the extinction or risk of extinction of wild species, including wild relatives of rice and cotton (Ellstrand *et al.*, 1999).
- A related concern is the possible introgression of transgenes into related native species in centres of origin or diversity (Box 1.1) leading to loss of potential breeding material.

An important question to ask is whether transgenic plants have a greater outcrossing potential than conventional plants. In an early study, it was shown that the outcrossing potential of a GM herbicide-tolerant *Arabidopsis* variety was approximately 20-fold higher than that of herbicide-tolerant *Arabidopsis* produced by chemical mutagenesis (Bergelson *et al.*, 1998), suggesting that transgenic crop varieties might have a greater propensity to outcross. However, subsequent failures to demonstrate the generality of this result indicate that other factors may have biased the system; in any case, the underlying mechanism resulting in this difference remains unknown.

B. Potential Adverse Impacts from Unintended Changes in the GM Plant

1. Potential Increase in the Weediness of the Crop Plant

Unintended changes in the GM plant that occur during the transformation process may increase the tendency of the GM plant to become a weed in agricultural systems or increase its survival in unmanaged systems. Seed may escape from transgenic crop harvests giving rise to volunteers (see Appendix A) in rotational agricultural systems in subsequent years. Plant volunteers are known to occur in conventional farming systems and, by extension, similar situations are possible in transgenic agriculture, particularly in those cases where the untransformed crop is known to be a volunteer problem. The biological basis of crop volunteerism is not well understood but it is likely that environmental factors induce increased seed shattering (uncontrolled dispersal of seed at maturity rather than retaining seed for harvesting) and dormancy leading to enhanced seed persistence. In addition, seed dispersal outside farming systems is also possible by means of wind, water and animal vectors or through the inadvertent spillage of seed during transportation.

However, predicting the invasiveness potential of a plant is not easy as it is a function of both biological and environmental factors. Most crop plants are unable to survive and persist outside of agricultural systems precisely because of the traits that have been selected during the domestication process. However, those that survive do so because they have characteristics that can lead to weediness (Box 4.8).

BOX 4.8

WEEDINESS*

The term weed is used to describe a plant that is a nuisance in managed ecosystems such as farms or forest plantations. Typically, weeds are plant species that spread easily in disturbed areas or among crops. "Weediness" potential is really a measure of a plant's ability to successfully colonize an ecosystem, especially when it may also lead to the displacement of other species. The main characteristics of weeds include:

- Discontinuous germination and long-lived seeds;
- Rapid seedling growth;
- Rapid growth to reproductive stage;
- Long continuous seed production;
- Self-compatible, but not obligatorily self-pollinated (see Box 4.10) or apomictic (see Appendix A);
- If outcrossing, using wind or an unspecialized pollinator (see Box 4.10);
- High seed output under favourable conditions;

---

*Based on http://www.agbios.com/cstudies.php?book=ESA&ev=MON810&chapter=Weediness&lang=

(cont'd)

BOX 4.8 (*cont'd*)

- Germination and seed production under a wide range of environmental conditions;
- High tolerance of climatic and edaphic variation (see Appendix A);
- Special adaptations for dispersal;
- High competitiveness achieved through, for example, allelochemicals (see Appendix A) or choking growth;
- If perennial, then with vigorous vegetative reproduction, brittleness at the lower nodes or of rhizomes or rootstocks, and ability to regenerate from severed rootstocks.

Generally, weediness depends on the selective advantage of many genes functioning in combination, which are unrelated to the genes usually introduced for agronomic reasons. However, traits which enhance tolerance to environmental stresses, such as drought, cold or dormancy, have the potential to increase the survival and distribution of the plant in managed and unmanaged ecosystems. Additionally, traits which provide for resistance to biotic stresses that play a significant role in the ecology of the plant (e.g., insect or pathogen resistance) could permit the plant to become persistent and/or invasive within and outside of the agricultural ecosystem. To evaluate if a transgenic plant has altered weediness potential in comparison with its conventional counterpart, the following may be examined:

- Dissemination of seed;
- Dormancy of seed;
- Germination of seed/survival;
- Competitiveness;
- Agronomic characteristics, for example time to maturity, disease and pest resistance;
- Stress tolerance.

Thus, GM plants can be assessed for these characteristics to ensure that they are unlikely to become weeds. Useful data for the environmental risk assessment includes the phenotype of the GM plant compared to that of a non-GM near isoline (see Appendix A) studied in a variety of environments. Typically, measurements should include morphological and developmental parameters, yield, and pest and disease incidence.

Studies conducted thus far indicate that GM plants do not have a fitness advantage in comparison to conventional ones and may even have reduced fitness (Marvier, 2001). Furthermore, a comparative study over a ten-year period involving herbicide-tolerant and insect-resistant GM and non-GM oilseed rape, potato, sugar beet and corn in 12 different habitats showed the GM varieties to be no more persistent or invasive than the non-GM ones (Crawley *et al.*, 2001).

2. *Indirect Adverse Impacts from Changes in Agricultural Practices*

As mentioned in Chapter 1, introducing new agricultural technologies brings with it other changes in agricultural practice, both because older practices are directly replaced (e.g., insecticidal GM plants versus conventional insecticides) and because the new technology facilitates other changes (e.g., changes in tillage associated with herbicide-tolerant GM plants). In this case, the potential hazards and their consequences are difficult to predict. Furthermore, these sorts of impacts can only be identified and accurately measured once the technologies are in commercial use at large scales. Therefore, these concerns are generally addressed through post-commercial monitoring of GM crops (discussed in Section VI) and systems-level comparisons of GM and non-GM crops.

It is important to note that agriculture is continuously changing because of technological developments, economic circumstances and environmental shifts. Therefore, detecting changes, and even apparently adverse changes, in agriculture is not surprising, but definitively assigning a cause may be difficult if not impossible.

A report by the US National Research Council (NRC, 2002) concluded that both GM and conventional methods of crop improvement can result in unintended effects on crop traits but that "the transgenic process presents no new categories of risk compared to conventional methods of crop improvement, but specific traits introduced by both approaches can pose unique risks". The implications of this are two-fold; first, that risk assessment should proceed on a case-by-case basis and second, that regulatory oversight should be triggered by the "novelty" of a given product rather than by the method by which it has been produced.

Changes in agronomic techniques to take advantage of a trait can even reduce problems that arise from conventional agriculture. For example, increased adoption of no- or low-tillage practices with herbicide-tolerant crops (Box 4.9) reduces the soil damage and erosion caused by conventional ploughing.

Some of the possible hazards that could arise from the deployment of GM crops include intensification of agriculture leading to monocultures that could encroach on natural habitats and reduced on-farm biodiversity through effects on non-target organisms mentioned above. However, the same risks are present with conventional agricultural practices.

It should be remembered that the impacts of new technologies on agricultural systems will be cumulative in some cases, or may simply take a number of years to be realized. Thus it probably will take several more years before the impact of GM crops on agricultural systems can be fully assessed.

BOX 4.9

NO AND LOW TILLAGE

Herbicide-tolerant crops have facilitated reductions in the amount of tillage because of the improved and more flexible weed control that they provide. In the USA and Argentina, the use of herbicide-tolerant soybean has been accompanied by a major reduction in tillage. Similar effects have been seen with herbicide-tolerant cotton in the USA. Reductions in tillage (see Table 1) bring many environmental benefits, including reducing soil erosion and runoff, maintaining soil fertility, and promoting in-field biodiversity.

TABLE 1   Benefits of reduced tillage.

Variable	Conversion to no till
Soil erosion reduction	>90%
Organic matter increase	0.1%/year
Increase in stored soil water	30%
Water infiltration	17 cm/hr
Increase in earthworms	275,000/hectare
Soil temperature reduction	9%
Decrease in evaporation	80%
Improvement in soil tilth	120%
Reduction in water runoff	70%

There is also an economic benefit from reducing tillage as the total cost of erosion in the USA is estimated to be $15–$44 billion (1995 figures). This value comprises on-site losses (e.g., nutrient loss, lower yields, degradation of soil quality) and off-site losses (e.g., destruction of aquatic habitat, respiratory health problems, increase in water treatment and dredging of lakes, rivers and waterways).

IV. EXPOSURE

A. Direct Exposure to the GM Protein and Plant

It is necessary to understand how and to what extent non-target animals and plants may be exposed to the novel protein(s) and the GM crop. We can use what is known about the biology and environment of the non-GM crop as a guide to understanding potential and actual exposure.

Various possible routes of exposure can be identified based on crop biology and the level of exposure to the novel protein and/or GM plant should be estimated for these different routes based on protein expression measurements (or presence) in target plant tissues under a range of environmental conditions. These routes of exposure include:

- Direct feeding on GM plants, as in the case of herbivores;
- Incidental feeding on GM pollen or other plant tissues that are moved by wind and rain, as in the cases of Monarch butterflies (see Box 4.5);
- Exudation of proteins into the soil, potentially affecting soil organisms;
- Secondary exposure as in the case of predators feeding on herbivores that had fed on GM plants;
- Movement of genes through pollen flow with subsequent expression of the GM trait in a new variety or a closely related species.

The level of exposure will depend upon the scale of release, the biology of the crop, and the nature of agricultural practices. Growing a GM crop in unconfined conditions, and in a variety of environments, obviously will lead to greater environmental exposure than confined field trials. Likewise, as discussed earlier, environmental exposure will be greater where inter-specific gene flow can occur than where it is impossible or managed.

B. Exposure to the GM Protein or Plant Through Gene Flow

Pollen-mediated gene flow is the primary gene flow mechanism for most crop plant species. Of paramount importance in assessing pollen-mediated gene flow is knowledge of the pollination mechanism (Box 4.10).

Pollen-mediated gene flow assumes importance if transgenes become established in viable hybrids and if the latter themselves become the risk source. The probability of cross-pollination of transgenic crops with sexually compatible relatives depends on:

- The pollen transfer potential of the GM crop. This depends to a large extent on the reproductive biology of the crop. Self-incompatible and dioecious plants (see Box 4.10) would be expected to have higher outcrossing potential than self-fertilizing species. Factors which have a major influence on pollen transfer potential include the characteristics of the pollen source (the plant and flower density and plant distribution), the size of the exposed population, the means of pollen dispersal (i.e., wind or animal vectors), and the viability of pollen in transit.
- The receptor plant. The size of the exposed population and synchronization of flowering with that of the transgenic crop determines to a large extent the cross-fertilization ability of the receptor plant. If the receptor plant happens to be a root or leaf crop, the harvested material

> ## BOX 4.10
>
> ## POLLINATION MECHANISMS
>
> There are two major types of pollination mechanism, namely self-pollination and cross-pollination.
>
> **Self-pollination**: the pollen fertilizes the female sex cell (ovum) of the same plant giving in-breeding (see Appendix A). Such plants have bisexual reproductive organs and are termed monoclinous or hermaphroditic.
>
> **Cross-pollination**: the pollen of one plant fertilizes the ovum of another plant of the same species giving outcrossing or outbreeding (see Appendix A). In cross-pollination, the pollen moves from plant to plant carried mainly by wind for some crops or mainly by insects, often bee species, for others. The success of cross-pollination depends on the synchrony of flowering, the longevity of pollen, and sexual compatibility. Cross-pollinating plants often have the male and female reproductive units on separate plants (termed dioecious). If they are on the same plant, there is self-incompatibility in which the pollen from that plant will not fertilize the female reproductive unit on that plant. With some outcrossing plant species (e.g., apple and cherry), there is interspecific incompatibility between certain varieties but other varieties are compatible. There is usually incompatibility between different species.
>
> Examples of pollination types of major crops:
>
> - Self-pollinated: cereals (barley, oats, rice, sorghum*, wheat), peanut, soybean, pea, cowpea, cotton*, flax, pepper, tomato, tobacco;
> - Cross-pollinated (wind): corn, sugarcane, sorghum*, oil-seed rape, beet;
> - Cross-pollinated (insect): avocado, alfalfa, onion, cotton*.
>
> Although plant species usually have one of the two types of pollination mechanisms predominating, they do not have it in 100% of the cases. For example, rice is self-pollinated but will cross-pollinate up to 5% of the time depending on variety. The type of pollination mechanism is extremely important with regard to environmental risk assessment for GM crops.
>
> _____
>
> *in-/outbred plants which use both self- and cross-pollination.

from F1 generation hybrids will not carry the transgene(s) and cross-pollination becomes an issue only if cross-pollinated seeds contaminate other crop plots. Also to be considered is the propensity of the receptor plant to show weediness characteristics (see Box 4.8).
- The probability of hybrid formation. This depends on the distribution of weedy or wild relatives and the internal barriers to hybrid formation. Pre-mating barriers to hybrid formation can be spatial,

determined by ecology, or reproductive, determined by temporal and floral divergence. Post-mating barriers to hybrid formation can be pre-zygotic (see Appendix A), for example incompatibility between pollen and pistil, or post-zygotic (see Appendix A), such as hybrid lethality, sterility or breakdown.

• The probability of hybrid survival. Hybrid survival can be compromised by genome, genome–cytoplasm and embryo–endosperm incompatibilities in the developing zygote.

Dispersal of seeds in natural and semi-natural habitats may also give rise to the establishment of persistent populations of individuals of a cultivated crop and subsequent hybridization with wild relatives. This may lead to the rapid evolution of viable hybrids, known as feral, which have lost traits associated with domestication. Although the biological basis of ferality is uncertain, it has been argued that novel traits conferring a selective advantage may contribute to the establishment of feral populations (Gressel, 2005). Whether or not GM could facilitate ferality is subject to speculation. Each wild-weed-crop complex needs to be studied on a case-by-case basis because of differences in the biological or environmental factors that contribute to domestication and de-domestication, their distribution in relation to sexually compatible wild plants and weeds, and the cultural practices in different farming systems (Bagavathiannan and Van Acker, 2008).

V. RISK MANAGEMENT OPTIONS

As described in Chapter 2 risk management can be proactive and preventive, or reactive.

Proactive risk management occurs at various points during the process of GM product design. In particular, choosing transgenes that have superior environmental safety profiles (such as Bt proteins for insect control) can reduce many concerns. Similarly, including genetic mechanisms that limit or preclude gene flow can be highly effective in managing potential environmental risks.

In addition, many potential environmental risks can be managed quantitatively through spatial or temporal isolation, or other forms of confinement. Risks stemming from gene flow can be minimized or prevented in this way. Other risk management applications include restricting the use of a GM product from certain regions where wild relatives or potentially vulnerable threatened or endangered species are present.

One special case of management for GM crops involves insect resistance to insecticidal Bt crops. These management programs are described in Box 4.11.

BOX 4.11

MANAGING INSECT RESISTANCE TO TRANSGENIC BT PROTEINS

Experience over many years has shown that prolonged exposure to a pesticide will select intensively for variants in the target pest population with resistance to that pesticide. Thus, there is concern that the widespread deployment of insect resistant crops expressing the Bt toxin would lead to resistance in the target insect and thus breakdown of the protection. Various insect resistance management (IRM) strategies have been suggested to minimize this problem including:

- Growing non-GM plants (termed refugia or refuges) adjacent to GM crops expressing the Cry protein at a high level (see Box 4.3 for a description of Cry proteins);
- Co-expression of different *cry* genes in the same plant.

The theory behind refugia is that the number of resistant pest insects surviving on the GM crop will be small and that these insects will mate with non-resistant insects coming from the refuge. As most resistance genes are recessive, the progeny of resistant insects from the Bt crop and susceptible insects from the refuge will not be resistant. Various studies have been undertaken to verify the assumptions behind this strategy: that resistance genes are naturally rare, that resistance genes tend to be largely recessive, and that insects from the GM crop will interbreed with those from the refuge (see Fig. 1).

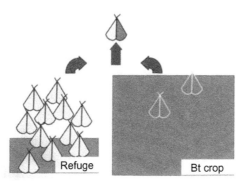

FIGURE 1   Refugia strategy. Resistant insects (red wings) in the Bt crop mate with the susceptible insects (yellow wings) in the refuge to give heterozygous progeny. The progeny from any insects from the Bt crop with a recessive resistance gene (the usual situation) will be susceptible (see colour section).

(cont'd)

BOX 4.11   *(cont'd)*

Typically, farmers growing Bt crops have been required to set aside about 20% of their farms to non-Bt crop varieties that act as refuge. After 12 years of use of Bt crops, it is generally felt that the global application of the refuge strategy for IRM has significantly delayed the evolution of resistance to Bt crops. However, there is some variation in success depending on the target species and the Bt protein used (Tabashnik *et al.*, 2008).

In addition, empirical and simulation modelling approaches have been used to understand what factors most affect the risk of resistance to Bt crops and the size of the refuge that is appropriate (e.g., Roush, 1998; Tabashnik *et al.*, 2003), including:

- If the Bt crop provides very high levels of control of the pest insects ("high dose"), the resistance risk is lower and less refuge is needed than for Bt crops that provide only partial control of the pest species;
- If the pest insects feed on many different crops, these alternative hosts may act as a natural source of refuge and reduce the risk of resistance;
- When two different, individually effective Bt proteins are combined in a single GM plant, the risk of resistance is dramatically reduced and the size of refuge needed is much smaller than for Bt crops with a single Bt toxin.

This last factor has driven the replacement of the first generation of Bt crops that contained a single insecticidal Bt protein by a second generation of products containing two different Bt proteins for control of the same target pests. The introduction of this second generation of Bt crops should reduce the risk of resistance evolving to the Bt proteins present in these GM products.

VI. MONITORING

Monitoring is a necessary complement to environmental risk assessment and risk management procedures, ensuring that the results of the risk assessment can be validated, uncertainties addressed and identified risks tracked. As described in Chapter 2, monitoring can be derived from risk assessment and be case specific, or more generally it can address potential unexpected effects on the agricultural system (surveillance).

Post-commercial monitoring may include elements of biodiversity that could not be adequately assessed at earlier stages of product development. For GM Bt crops, monitoring also includes resistance monitoring to ensure that resistance management programmes are functioning appropriately.

This case-specific monitoring is typically carried by the product registrants, with regular reporting to regulatory authorities, and/or by governmental research laboratories where the capacity exists. This routine monitoring may be supplemented by government-funded research programmes that address particular concerns or side issues in more detail (for example, USDA-funded grants in the USA).

In addition, approaches for general surveillance have been developed by regulatory agencies such as the European Food Safety Authority (EFSA, 2006) involving regular interviews of farmers growing GM crops to assess the net positive and negative impacts resulting from introducing a particular GM crop. These surveys also seek to understand whether agricultural practices are being affected indirectly.

References

Bagavathiannan, M.V. and Van Acker, R.C. (2008). Crop ferality: implications for novel trait confinement. *Agriculture, Ecosystems and Environment* **127**, 1–6.

Bergelson, J., Purrington, C.B. and Wichmann, G. (1998). Promiscuity in transgenic plants. *Nature* **395**, 25.

Christou, P. (2001). No credible evidence is presented to support claims that transgenic DNA was introgressed into traditional maize landraces in Oaxaca, Mexico. *Transgen. Res.* **11**, 3–5.

Crawley, M.J., Brown, S.L., Hails, R.S., Kohn, D.D. and Rees, M. (2001). Transgenic crops in natural habitats. *Nature* **409**, 682–683.

Eastham, K. and Sweet, J. (2002). Genetically modified organisms: the significance of gene flow through pollen transfer. *Environmental Issues Report* **28**, 1–75.

EFSA (2006). Opinion of the Scientific Panel on Genetically Modified Organisms on the Post Market Environmental Monitoring (PMEM) of genetically modified plants. *The EFSA Journal* **319**, 1–27.

Ellstrand, N.C. (2003). *Dangerous Liaisons? When Cultivated Plants Mate with Their Wild Relatives.* The Johns Hopkins University Press, Baltimore, MD.

Ellstrand, N.C., Prentice, H.C. and Hancock, J.F. (1999). Gene flow and introgression from domesticated plants into their wild relatives. *Annu. Rev. Ecol. Syst.* **30**, 539–563.

EPA (2007). White Paper on Tier-Based Testing for the Effects of Proteinaceous Insecticidal Plant-Incorporated Protectants on Non-Target Arthropods for Regulatory Risk Assessments. http://www.epa.gov/pesticides/biopesticides/pips/non-target-arthropods.pdf.

Gressel, J. (ed.) (2005). *Crop Ferality and Volunteerism: A Threat to Food Security in the Transgenic Era?* CRC Press.

Hall, L., Topinka, K., Huffman, J., Davis, L. and Good, A. (2000). Pollen flow between herbicide-resistant Brassica napus is the cause of multiple-resistance B. napus volunteers. *Weed Sci.* **48**, 688–694.

Losey, J.E., Raynor, L.S. and Carter, M.E. (1999). Transgenic pollen harms Monarch larvae. *Nature* **399**, 214.

Marvier, M. (2001). Can risk analysis "colorise" the black and white of transgenic crops? *Plant Health Prog.* Accessible at: http://www.aspnet.org/online/feature/riskanalysis/

NRC (2002). *Environmental Effects of Transgenic Plants: The Scope and Adequacy of Regulation.* National Research Council, National Academies Press, Washington, DC.

Romeis, J., Bartsch, D., Bigler, F., Candolfi, M.P., Gielkens, M.M.C., Hartley, S.E., Hellmich, R.L., Huesing, J.E., Jepson, P.C., Layton, R., Quemada, H., Raybould, A., Rose, R.I., Schiemann, J., Sears, M.K., Shelton, A.M., Sweet, J., Vaituzis, Z. and Wolt, J.D. (2008).

Assessment of risk of insect-resistant transgenic crops to nontarget arthropods. *Nature Biotechnol.* **26**, 203–208.

Roush, R.T. (1998). Two toxin strategies for management of insecticidal transgenic crops: can pyramiding succeed where pesticide mixtures have not? *Phil. Trans. R. Soc. Lond.* **353**, 1777–1786.

Quist, D. and Chapela, I.H. (2001). Transgenic DNA introgressed into traditional maize landraces in Oaxaca, Mexico. *Nature* **414**, 541–543.

Sears, M.K., Hellmich, R.L., Stanley-Horn, D.E., Oberhauser, K.S., Pleasants, J.M., Mattila, H.R., Siegfried, B.D. and Dively, G.P. (2001). Impact of Bt corn pollen on monarch butterfly populations: a risk assessment. *Proc. Natl. Acad. Sci. USA* **98**, 11937–11942.

Snow, A.A., Uthus, K.L. and Culley, T.M. (2001). Fitness of hybrids between weedy and cultivated radish: implications for weed evolution. *Ecologic. App.* **11**, 934–943.

Tabashnik, B.E., Carrière, Y., Dennehy, T.J., Morin, S., Sisterson, M.S., Roush, R.T., Shelton, A.M. and Zhao, J.Z. (2003). Insect resistance to transgenic Bt crops: lessons from the laboratory and field. *J. Econ. Entomol.* **96**, 1031–1038.

Tabashnik, B.E., Gassmann, A.J., Crowder, D.W. and Carrière, Y. (2008). Insect resistance to *Bt* crops: evidence versus theory. *Nature Biotechnol.* **26**, 199–202.

Wisniewski, J.P., Frangne, N., Massonneau, A. and Dumas, C. (2002). Between myth and reality: genetically modified maize, an example of a sizeable scientific controversy. *Biochimie* **84**, 1095–1103.

Wolfenbarger, L.L., Naranjo, S.E., Lundgren, J.G., Bitzer, R.J. and Watrud, L.S. (2008). Bt crop effects on functional guilds of non-target arthropods: a meta-analysis. *PLoS ONE* **3**, e2118 (available at www.plosone.org).

Further Reading

Bravo, A. and Soberón, M. (2008). How to cope with insect resistance to Bt toxins. *Trends Biotechnol.* **26**, 573–579.

Environmental Biosafety Research. www.ebr.journal.org.

Risk Perception and Public Attitudes to GM

ABSTRACT

As well as using scientific evidence, decision makers have to consider other factors such as the opinions of different stakeholders and the balance between costs and benefits of either approving or not approving new technologies. This chapter discusses the factors involved in shaping stakeholder opinions, and how and what information on GM products is communicated.

I. INTRODUCTION

Chapter 2 makes clear that the decision-making process on GMO safety takes into account not only scientific assessment but also considers other aspects such as trade implications, the balance of risks against benefits, and public opinion. Thus, the decision-making process receives diverse inputs and feedback from policy makers, the scientific community, the agricultural sector and public interest groups.

Public perception of risk in biotechnology is influenced to a large extent by the effectiveness of regulation, the prevailing ethical climate, and broader economic and social considerations. Equally, biosafety and trade regulation are shaped by public perception.

This chapter discusses these inputs and especially factors involved in shaping public opinion. For this discussion, the public is defined as the end user of the GM technology or the consumer of its products. The uptake of any technology is dependent on public attitudes towards it. As GM technology is relatively new, consumers need to be informed about the pros and cons of it; consumer attitudes will also be affected by their ability to identify GM products. Therefore the labelling of these products is potentially important and different approaches to labelling will be described.

Other socio-economic implications from the adoption of GM plant technology are equally important in decision making but are beyond the scope of this book. In-depth discussion of such implications is given by a number of papers some of which are listed in the References section.

II. PERCEPTION OF RISK

A. Introduction

Chapters 2, 3 and 4 emphasized that the risk assessment process is science based but involves an inherent amount of scientific uncertainty as mentioned in Chapter 2. Risk perception among scientists varies, with reductionist scientists tending to be less risk averse than scientists who take a systems approach. Likewise, risk perception among farmers varies too, but the uptake of GM crops worldwide (Chapter 1, Section IV.A) indicates that many farmers consider that the benefits from the technology far outweigh the costs (potential risks). However, in countries where organic farming constitutes a sizeable economic activity, there may be economic concerns because of potential "contamination" of organic crops by gene flow. The rules for organic farming in many countries make the use of GM technology difficult. This raises the issue of coexistence of crops grown in different agricultural systems, and of liability and redress which will be discussed in Section VI.

B. Factors Influencing Public Attitudes Toward GM Technology

Numerous other factors influence public attitudes toward GM technology over and above scientific uncertainty and farmer preferences. These factors raise issues that are inherently socio-economic in character and cannot be addressed by scientific research. They include:

1. **Perception of risk** (is the technology safe?). Risk is one of the complex indicators of human/environment interactions and the assessment of risk is a continuous activity for each human being. In many countries, public opinion is a major factor in the decision-making process for the release of GM crops to the environment and the use of GM products in food, although this may depend upon national features such as the political system, the media and the level of education. Thus, it is important to understand the factors that influence public perception of risk (Box 5.1).

2. **Ethics** (does the science "meddle" with nature?). As noted in Box 5.1, a "natural" product is often regarded as less risky than one in which "man has interfered". Coupled with this is the uncertainty about products resulting from a new, unfamiliar technology which bypasses natural constraints. Such ethical considerations vary among cultures and religions and are discussed in depth in Bruce and Bruce (1998).

3. **Economics and politics** (what are the impacts of the technology on the national and personal economy and on food security?). The introduction of a new technology may have both positive and negative impacts on national and personal economies. It may open up new economic opportunities but may also affect trade (see Chapter 6) and lead to labour displacement and labour disputes. It may also help to increase agricultural productivity and thus national food security and decrease the cost of food for consumers or, alternatively, may have a negative impact on food security through the adoption of monocultures.

BOX 5.1

FACTORS INVOLVED IN SHAPING THE PUBLIC'S PERCEPTION OF RISK

Several studies on the factors which impact on the public's perception of risk identified the following main points:

1. Familiarity

Unfamiliar or novel risks cause people to worry more. The sense of risk rises if it is believed that an activity or technology is not well understood. Knowing a victim raises the sense of risk. Bad events in the past raise the sense of risk.

2. Trust

Concern is greater when the people who are supposed to protect us or those exposing us to the risk or communicating about it are less trusted.

3. Scale of risk

The scale of a risk affects the perception of it. A catastrophic event in which fatalities occur in large numbers, such as an airplane crash, leads to the risk (of flying) being considered to be greater than that of a single event (travelling by car), even though overall there are more deaths caused by the latter. Similarly, a risk that kills you in a dreadful way evokes more fear than one that kills you benignly.

4. Source of risk (natural or man-made)

Man-made risks are perceived to be greater than natural ones.

5. Choice

Awareness of risk allows one to select among different options.

6. Children and future generations (short term v. long term)

A risk appears to be worse if children or future generations are involved. However, more immediate threats are often more important whereas those in the future tend to be discounted.

7. Benefits

If the benefits of an activity or technology are not clear, the risk is perceived to be greater.

8. Reversibility

If the effects of something going wrong can be reversed, the perception of risk decreases. People are more concerned if they feel that the potential for harm is beyond their control.

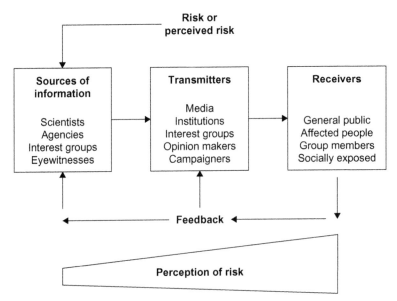

FIGURE 5.1   Social amplification of risk. Diagram showing how the perception of risk is amplified. Modified from Pidgeon *et al.* (2003).

4. Culture (what are the changes to traditional production systems?). In many countries, traditional food production systems are an inherent part of the culture. For instance, the MILPA crop growing system throughout Mesoamerica is based on the ancient agricultural methods of the Mayan people. It is an integrated system for the production of corn, beans, lima beans and squash. In many countries there are personal preferences for specific foods or crop varieties of . For example, the main type of corn consumed in the USA is yellow whereas in Mexico and South Africa it is white. In Asia, different nationalities have preferences for different traditional varieties of rice, but there also is a move towards higher income Asian countries and non-rice producing countries preferring convenient (easily cooked) varieties (Suwannaporn *et al.*, 2008).

5. Social amplification of risk. Social dynamics influence how risks are represented and communicated. Risk events have a "signal value" that is propagated through a social network (Fig. 5.1). Thus, the issues arising from a risk associated with a particular hazard can be amplified (or decreased) during the direct flow and feedback of information from the source to the receivers. The input from the sources and transmitters has a considerable influence on the perception of a risk by the receivers.

C. Public Perception of GM Products in Food

There have been numerous studies in many countries on consumers' perception of the relative hazards in food; some examples are given in Table 5.1.

The Eurobarometer is the longest-running set of surveys into consumers' perceptions, some of the major findings related to GM food being (Gaskell *et al.*, 2003, 2006):

- Optimism about biotechnology in general increased in the period 1999–2005 after decreasing from 1991–1999.
- Europeans distinguish between different types of biotechnology applications, particularly medical in contrast to agri-food.
- The majority of Europeans do not support GM foods because they are judged not to be useful, to be morally unacceptable, and to be risky for society. Unless new crops are seen to have consumer benefits, the public may continue to be sceptical.
- Europeans are concerned about the fragility of nature and about the impact of human actions and technology upon nature.
- The "engaged" (more aware and knowledgeable) are more supportive of biotechnology than the less engaged.

TABLE 5.1   Examples of surveys on public perception of biotechnology and GM.

Country	Source	Method	Year	Main conclusions
Europe EU	Eurobarometer	Surveys	Frequently	Differences in different countries
UK	GM Nation	Debates	2004	See text
N. America USA	Hoban (2004)	Surveys	1992–2003	Lack of awareness of GM content of food Not much concern about GM food
Asia, China	Hoban (2004)	Surveys	2002	Many consumers aware of GM in food
Indonesia, the Philippines				Most consumers not concerned
Australasia Australia	Cormick (2003)	Focus groups	2001	Confusion over GM Increasing number of people willing to eat GM food
New Zealand	Hunt *et al.* (2003)	Focus groups on specific exemplars	2003	Acceptability on a case-by-case basis Need for more information Ethical issues

- Men tend to be more supportive of biotechnology than women and 15–39-year-old people more than those over 55 years old.
- About 70% of Europeans have confidence in doctors, university scientists and consumer and patients' organizations. About 55% have confidence in scientists working in industry, the media, environment groups, shops farmers and the European Commission. Fewer than 50% have confidence in their own governments and in industry.
- The majority of Europeans accept the principle of decision making based on scientific evidence rather than on moral and ethical criteria.

The Eurobarometer surveys do not reflect public attitudes in other parts of the world but they illustrate the complexity of factors that affect opinions of biotechnology. Among the many factors in this complex situation are the attitude to risk (see Box 5.1) which will be affected by the level of knowledge and familiarity with the subject (as noted in Box 5.2,

---

BOX 5.2

STUDY ON "DO EUROPEAN CONSUMERS BUY GM FOOD?"

This study was a project funded under the European Commission Framework Programme 6 and the final report was published in October 2008 (http://www.kcl.ac.uk/consumerchoice).

Surveys of retailers were undertaken in ten EU countries to determine which GM-labelled and GM-free-labelled products were for sale, and of consumers to assess what they had bought and what they would prefer. The main results were:

- Most shoppers do not actively avoid GM-labelled products.
- The majority of consumers pay scant attention to labels. When they do, labels are interpreted as warning rather than simply as information.
- Responses given by consumers when prompted by questionnaires about GM foods are not a reliable guide to what they do when shopping.
- Consumers pointed to perceived environmental and health risks but were generally less aware of possible benefits than of potential hazards.
- A major factor in governing the purchase of GM products by Europeans is the decision of retailers to make them available to consumers.
- Focus groups showed that (a) GM food is not a topic at the forefront of consumers' minds when discussing food purchasing habits; (b) labelling was regarded as important but few actually looked at labels; (c) consumers did not appear to be well informed about genetic modification and had not given it much thought.

familiarity is one of the factors involved in risk perception), the concept of "naturalness" (Madsen *et al.*, 2002), and the perception of who benefits from the technology. The weight that these and other factors have on public perception will vary among countries, cultures and experiences. For instance, the coincidence of the arrival of GM products with two unrelated health scares (contaminated blood and mad cow disease) is suggested to have contributed to the attitude in Europe.

As public opinion is often changing due to factors such as increases in information, adverse publicity and economic circumstances making GM products financially more attractive, one-off surveys can quickly become out of date. Furthermore, how the surveys are conducted can influence the results. The wording of question(s) can affect the response as the focus on the technology and its products is likely to elicit a social amplification of risk. Three examples illustrate the problems with obtaining meaningful information from surveys:

1. In a recent report from the UK Cabinet Office, public concern for GM food ranked relatively low among perceived food hazards and has been declining between 2001 and 2008 (Fig. 5.2). This differs somewhat when the emphasis of the survey is on the acceptability of GM products in food. A similar situation has been found in the USA (IFIC, 2008).
2. A nationwide debate, termed GM Nation, was held in the UK in 2004 and comprised 600 meetings held in different parts of the country (details can be found at http://www2.aebc.gov.uk/aebc/reports/gm_nation_report_final.pdf). The debate was welcomed and there was a broad desire to know more and for more research to be done, but people were generally uneasy about GM and the more people

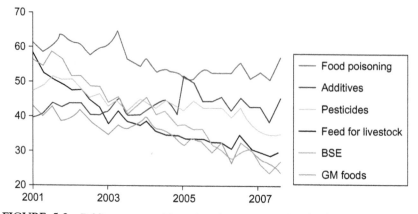

FIGURE 5.2   Public concern with various issues concerning food. From UK Cabinet Office (2008). With kind permission of the UK Cabinet Office (see colour section).

engaged in GM issues, the harder their attitudes and more intense their concerns.

3. A recent large study within the EU (Box 5.2) attempted to avoid the problem of loaded questions and showed that consumer attitudes to GM food are complex.

Although the above examples mostly relate to the EU, they illustrate some of the more general problems in undertaking and interpreting surveys on attitudes to GM products. Obviously there are huge differences in the public perception of this technology among regions, and even within a region, and these will also vary over time. For a more detailed analysis of worldwide attitudes to agricultural biotechnology, see Hoban (1998, 2004) (Box 5.3).

BOX 5.3

SUMMARY OF THE POINTS IN THE SURVEYS REPORTED IN HOBAN (2004)

1. Over two-thirds of respondents in the USA, Colombia, Cuba, Dominican Republic, China, India, Indonesia and Thailand agreed that the benefits of GM crops are greater than the risks. On the other hand, fewer than 40% of consumers in Europe (France, Greece, Italy, Spain), Japan and South Korea saw the benefits as greater than the risks.

2. There clearly are differences in public acceptance of different biotechnological products. Survey participants were asked whether they would support or oppose the use of biotechnology to different applications. Eighty-five per cent of respondents indicated that they would support the use of biotechnology to develop new human medicines. However, 15% would oppose the use of biotechnology even for such a clearly beneficial use. About three-quarters of people reported support for environmental clean-up. Any mention of "animals" caused support to drop. Just over half (55%) expressed support for genetically modified animal feed (even when this resulted in healthier meat). Only 42% supported the use of biotechnology to clone animals for medical research.

3. Consumers in ten countries were asked whether they would buy food with GM ingredients if the resulting products were higher in nutrition. Respondents were given the option of continuing to buy the product or to stop buying it if they learned it was genetically modified. Consumers in China and India are clearly the most enthusiastic about these crops. There is also support among a majority of consumers from the USA, Brazil and Canada. On the other hand, a majority of European and

(cont'd)

BOX 5.3   (cont'd)

Australia consumers would tend to reject GM foods even if they were more nutritious.

4. For the first half of the 1990s, awareness of biotechnology in the USA remained rather low at about one-third. It hit a peak in 1997 when a survey was conducted soon after the news about Dolly, the cloned sheep. After that, awareness dropped until May 2000, then began rising again, reaching a peak of 53% awareness in June 2001. Since then awareness has dropped again as other issues dominated the public agenda in the USA.

5. US consumers were uncertain about the safety of GM foods. The perception of safety of GM food is dependent on the familiarity with the technology and its use in daily life. When asked initially, with little background information, whether GM foods are safe, almost half of the respondents said that they did not know, 29% said they are basically safe, and 25% said they are basically unsafe. However, after being informed that more than half of the products at the grocery store contain GM ingredients, almost half said that GM foods are safe, only 21% said that they are unsafe, and 31% said they were unsure. In fact, one in five of those who initially said GM foods were unsafe changed their minds. Thus, when some consumers learn how widespread GM foods are, they are more likely to believe they are safe.

6. The majority of Asian consumers were aware of the presence of biotechnology-derived foods in their everyday diets and were not worried about this situation. A majority of consumers reported that they believed they had eaten genetically modified foods, took no action to avoid such products and were willing to try samples of genetically modified foods.

7. More than 90% of Asian respondents reported personal concerns regarding nutrition and food safety. Those of greatest concern were nutritional value, microbial contamination and pesticide residues. GM foods were rated as the issue of least concern.

8. The food industry plays a vital role in shaping consumers' attitudes and appetite for new food items. This is particularly true for the products developed with biotechnology.

D. Public Perception of GM Crops and the Environment

There have been very few general studies on public attitudes to any impact that GM crops might have on the environment, particularly in comparison to the alternative practices discussed in Chapters 1 and 4.

The UK GM Nation debate, described above, covered potential environmental impacts. However, there are specific situations, such as the potential impact that transgenic maize may have in Mexico (Box 5.4), that have been analysed in detail.

BOX 5.4

ENVIRONMENTAL AND SOCIO-CULTURAL FACTORS ASSOCIATED WITH MAIZE IN MEXICO

A study on the potential impacts of the introduction of transgenic maize into Mexico (CEC, 2004) identified a number of biodiversity and socio-cultural factors that were particularly relevant for the release of transgenic maize in Mexico. These factors include the following.

Biodiversity

There are three areas of biodiversity that have special interest:

- The genetic diversity of maize and species of teosinte, all members of the genus *Zea*; Mexico is a Centre of Origin of maize (Box 1.1).
- The diverse biocoenoses of plants and animals that regularly occur in the fields where maize is cultivated.
- The biodiversity of neighbouring natural communities and ecosystems.

Socio-Cultural Issues

Maize has important cultural, symbolic and spiritual value for many Mexicans. Although its relative teosinte is considered by some to be a weed that reduces productivity, it is kept in milpas (local agronomic systems) in many areas because it is considered the "mother of maize". Teosinte is thereby a source of genetic variability for the different wild species of the genus *Zea* and for the planted landraces or varieties of maize.

Experimental planting and breeding of maize has been practised for several thousand years and is the starting point for the generation of the many native landraces of maize. Mexican landraces are neither genetically static nor genetically homogeneous; they are constantly being changed by those who use them. As part of this process, genes from improved/modern varieties are sometimes deliberately or inadvertently introduced into the landraces.

Campesinos (smallholder farmers with fewer than five hectares) regard freedom to exchange seed, to retain seed for future planting and experimentation with new seeds as fundamental to preservation of their landraces

(cont'd)

BOX 5.4   (*cont'd*)

and their cultural identity and communities. In general, there have not been formal systems among campesinos for *in situ* or *ex situ* conservation of landraces for the purpose of preserving genetic diversity. However, there are some formal systems among indigenous communities for *in situ* maintenance of specific maize varieties for cultivation and breeding.

A significant percentage of campesinos consider the presence of any transgenes in maize as an unacceptable risk to their traditional farming practices, and the cultural, symbolic and spiritual value of maize. That sense of harm is independent of its scientifically studied potential or the actual impact upon human health, genetic diversity and the environment. Furthermore, to many people in rural Mexico, the introgression of a transgene into maize is not acceptable and is considered "contamination".

III. INFORMATION AND PUBLIC TRUST

A. Introduction

In Chapter 2, we presented methodologies for conducting risk assessment. The initial and possibly the most important part of the risk assessment process is the development of a risk hypothesis. This, in turn, depends not only on empirical evidence but also on subjective assumptions regarding the nature and relative importance of scientific uncertainty. The justification of risk assumptions is essential in ensuring the acceptability of risk decisions. It has been argued that such justification cannot be provided by science alone but depends on factors such as (Economic and Social Research Council, 1999):

- The legitimacy of the institution making the justification;
- The degree of democratic accountability to which the institution is subjected; and
- The ethical acceptability of the assumptions adopted.

The above factors enhance public trust in regulatory institutions and greatly influence the perception of risks associated with GMOs. Public trust is further enhanced by the quality of information provided to the public, as well as its provenance. Academia, in general, is regarded as a trustworthy source of information.

Information relevant to GM technology reaches the consumer mainly through (i) primary sources that generate factual data and information,

TABLE 5.2   Trust in sources of information.

Trust	UK[a]	USA[b]
Most trust	Independent agency	Farmers[c]
	Independent scientist	Friends and relatives
	Consumer groups	Independent scientists
	Farmers	Food industry
	Government	Consumer groups
	Food industry	Independent agency
	Campaign groups	Campaign groups
	Media	Media
Least trust	Friends and relatives	Government

[a]Data from a survey by the Institute of Grocery Distribution in October 2008 (www.igd.com). The question asked was "Who would you trust most for accurate information about GM food?".
[b]Data from Ekanem et al. (2004). Ranking from % respondents who showed moderate and high trust.
[c]Extension professionals.

for example companies (technology developers), academia, and regulatory agencies, and (ii) secondary sources that proliferate and disseminate existing information, for example the media, non-governmental organizations and public advocacy groups. The major concerns about these sources of information are how much they can be trusted, how credible is the information that they supply, and how biased is that information. The results of surveys (Table 5.2) show that there are differences in the perception of trustworthiness of sources of information among countries. However, it is difficult to make detailed comparisons among independent surveys carried out at different times with different questions.

Some of the reasons why there is trust or distrust in sources of information are shown in Fig. 5.3.

What is clear is that, in most cases, the most trusted sources are those free from apparent commercial interest. Sources which are unlikely to present a balanced view are treated with more scepticism, for example campaign (activist) groups, the industry and media.

B. Sources of Information

Academia is generally regarded as a reliable source of information (Table 5.2). This is attested by surveys showing that academia enjoys considerable public authority and is trusted more than NGOs and other public interest groups (Aerni and Bernauer, 2006).

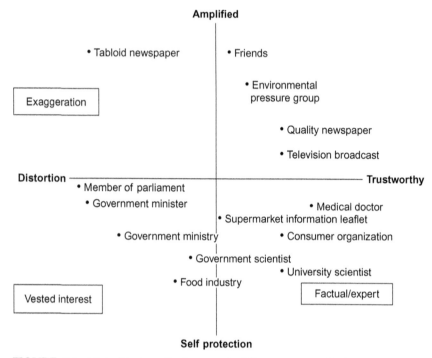

FIGURE 5.3   Plot of factors affecting trust in different sources of information on food safety. Adapted from Fig. 4 of Frewer *et al.* (1996) with kind permission of the publishers Wiley-Blackwell.

Increasingly, government agencies dealing with GMOs provide the general public with information relevant to the technology itself as well as aspects related to food and environmental safety. Arguably, such information may reflect broader governmental policies in the area of GMOs.

Other sources of information include private enterprises engaging in biotechnology R&D, as well as public interest groups. Ethical issues concerning GM technology have also been raised by religious and other organizations. Among the points of concern are the bypassing of the limitation on natural transfer of genes between sexually compatible species ("Who are we to act as God?") and the patenting of genes by biotechnology companies ("patenting life"). Arguments about "naturalness" are inherently complex and have been reviewed in depth elsewhere (Nuffield Council on Bioethics, 2006).

Although information from these sources may be perceived as biased, it is still useful to the "educated" reader in that it provides the complete spectrum of issues and arguments relevant to biotechnology and biosafety. Box 5.5 shows a list of some sources of information.

BOX 5.5

SOURCES OF INFORMATION ON GM ISSUES

(the list is indicative and is not intended to be comprehensive)

Sense About Science (www.senseaboutscience.org.uk)

Sense About Science is a UK-based independent charitable trust which responds to the misrepresentation of science and scientific evidence on issues that matter to society, from scares about plastic bottles, fluoride and the MMR vaccine to controversies about genetic modification, stem cell research and radiation. The organization works with scientists and civic groups to promote evidence and scientific reasoning in public discussion.

The Pew Initiative on Food and Biotechnology (http://www. pewtrusts.org/our_work_detail.aspx?id=442)

The Pew Charitable Trusts are USA based with comparable aims to Sense About Science. Pew applies an analytical approach to improve public policy and inform the public. In a series of reports, its Initiative on Food and Biotechnology focused on policy issues relating to GM food.

The International Service for the Acquisition of Agri-biotech Applications (ISAAA) (http://www.isaaa.org/kc/default.asp)

ISAAA is a not-for-profit organization that aims to deliver the benefits of new agricultural biotechnologies to the poor in developing countries. The Global Knowledge Center on Crop Biotechnology (KC) is the information-sharing initiative of ISAAA, and produces small booklets, termed Pocket Ks, on a wide range of biotechnology subjects.

The International Life Sciences Institute (ILSI) (http://www. ilsi.org/AboutILSI/ifbic.htm)

ILSI is a private-sector funded global network of scientists that focuses on enhancing the scientific basis for public health decision making. The ILSI International Food Biotechnology Committee (IFBiC) supports the development and harmonization of science-based regulations around the world for biotechnology-derived food products and disseminates science-based information regarding the safety assessment of these products to governments, industry, academia and other interested groups around the world.

(cont'd)

BOX 5.5   *(cont'd)*

The International Food Policy Research Institute (IFPRI) (http://www.ifpri.org)

IFPRI is one of 15 centres supported by the Consultative Group on International Agricultural Research, which is an alliance of 64 governments, private foundations and international and regional organizations. As part of its mission, it disseminates results of research that are critical inputs on food policy and the formulation of sound and appropriate policies.

EFSA (http://www.efsa.europa.eu/EFSA/efsa_locale-1178620753812_1178621185493.htm)

EFSA carries out the risk assessment of GMOs in the European Union. The website provides information for the general public on issues related to the safety of GM products.

GMO compass

General: http://www.gmo-compass.org/eng/home/ General information for the broader public on GMO issues. The site hosts a database of GM-derived products.

Environment: http://www.gmo-compass.org/eng/safety/environmental_safety/

Health: http://www.gmo-compass.org/eng/safety/human_health/

EC DG Health and Consumer Protection (http://ec.europa.eu/food/food/biotechnology/index_en.htm)

This site of the European Commission's DG Health and Environment hosts information on GMO issues as well as on relevant legal framework of the EU.

Food Standards Australia New Zealand (http://www.foodstandards.gov.au/foodmatters/gmfoods/ and http://www.foodstandards.gov.au/foodmatters/gmfoods/frequentlyaskedquest 3862.cfm)

The sites are hosted by the organization Food Standards Australia New Zealand (FSANZ). FSANZ is the competent authority to set food safety standards and to conduct GM food safety assessments on a case-by-case basis for Australia and New Zealand. Databases of current applications and approvals as well as related discussions are published on these sites.

(cont'd)

BOX 5.5   (cont'd)

Germany (http://www.gmo-safety.eu/en/)

This internet portal provides information about current and past bio-safety research into genetically modified plants in Germany. The informa-tion portal is designed to make research findings on the environmental safety of genetically modified plants accessible to the interested public and to contribute towards objective opinion forming.

UK Health and Safety Executive (http://www.hse.gov.uk/biosafety/gmo/whatare.htm)

The UK Health and Safety Executive (HSE) operates and enforces legis-lation to control the risks to human health and the environment related to the contained used of GMOS.

USDA (http://www.usda.gov/wps/portal/!ut/p/_s.7_0_A/7_0_1OB?contentidonly=true&navid=AGRICULTURE&contentid=BiotechnologyFAQs.xml)

The US Department of Agriculture (USDA) is one of the primary com-petent governmental agencies for regulating biotechnology in the USA. USDA regulates plant pests, plants and veterinary biologics regarding the safe to grow of plants. The agency has published a comprehensive FAQ list on biotechnology.

Singapore (http://www.gmac.gov.sg/)

The multi-agency Genetic Modification Advisory Committee (GMAC) was established under the purview of the Singaporean Ministry of Trade and Industry to oversee and provide scientifically sound advice on the research and development, production, release, use and handling of genetically modified organisms (GMOs) in Singapore. The site provides comprehensive information on GM technology, the regulatory process in Singapore as well as direction to further relevant resources.

WHO (http://www.who.int/foodsafety/publications/biotech/20questions/en/ and http://www.who.int/foodsafety/biotech/en/)

The World Health Organization (WHO) is the directing and coordi-nating authority for health within the United Nations system. WHO has been addressing a wide range of issues in the field of biotechnology and human health, including safety evaluation of vaccines produced using

(cont'd)

BOX 5.5   (*cont'd*)

biotechnology, human cloning and gene therapy. The organization provides useful information on food safety related to GM foods.

Europabio (http://www.europabio.org/bi_index.htm)

The European Association of Bioindustries (EuropaBio) publishes information material on current issues in agricultural, health and industrial biotechnology as well as biofuels.

Bio(http://bio.org/foodag/background/)

The US Biotechnology Industry Organization (BIO) provides background information on GM technology, current issues and regulatory procedures with a US focus.

Canada (http://www.hc-sc.gc.ca/fn-an/gmf-agm/index-eng.php)

Health Canada is the Federal department responsible for helping Canadians maintain and improve their health, while respecting individual choices and circumstances. This site hosts general information on GMOs (such as terminology, FAQs), an overview about the regulatory system in Canada as well as databases of current approvals.

IV. COMMUNICATING INFORMATION

A. Information

It is important that there is clear communication about any new technology like GM at a variety of levels and situations. In Chapter 2, we stressed that formal communication is essential among interested parties (stakeholders) in the risk assessment process. There also needs to be effective informal communication between experts and the lay public. Much of the discussion on GM technology has been highly polarized between advocates and opponents. Many decision makers and much of the general public are either swayed by the polarized arguments or confused about the issues.

A series of focus workshops (European Federation of Biotechnology, 2004) noted that the "deficit model" of communication, which directly relates public attitudes towards biotechnology to the availability of objective information, cannot be successful. They put forward an "interactive

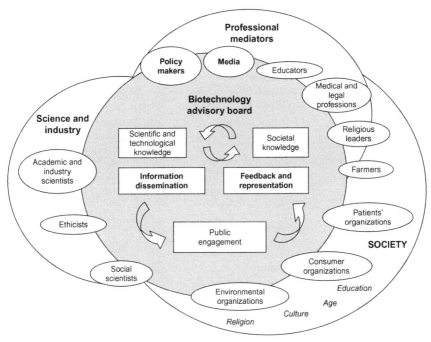

FIGURE 5.4 A model for biotechnology communication. From European Federation of Biotechnology (2004), with kind permission of the European Federation of Biotechnology.

model" shown in Fig. 5.4. This brings together all the groups and people involved in biotechnology.

There is no shortage of techniques developed by social scientists with a view to facilitating the discourse between experts and the lay public. Such techniques, sometimes called Deliberative and Inclusionary Processes, include consensus conferences, focus groups and surveys.

B. Communication Techniques

Some of these techniques are presented in Table 5.3.

Rather than representing a set of methodological rules, they represent a form of decision making where participants can exchange views among themselves and with decision makers. Participants may include scientists with expertise in plants, agriculture and molecular biology, the medical profession, the legal profession, religious leaders, ethicists and social scientists. It is imperative that any communication is a dialogue rather than being didactic so that the "expert" understands the concerns of the "non-expert" and, if necessary, discusses them in depth. This can raise difficulties as the basics can be lost in a fog of scientific terminology and detailed

TABLE 5.3   Deliberative and inclusionary processes (DIPs) (Economic and Social Research Council, 1999).

Method	Description	Output
Focus group	• 6–10 individuals • Discussion on set topics moderated by facilitator	• Record and analysis of public's main concerns
Citizens' juries	• 12–25 participants representative of the local community • Evaluation of decision alternatives • Witnesses present evidence (experts, scientists, local authority officers, politicians, pressure groups, business managers and indeed members of the public with relevant knowledge)	• Conflicts resolved • Decision by majority vote • Eventually new policy options developed
In-depth groups	• Like focus groups • Long-term process	• Record and analysis of public's main concerns
Consensus conferences	• Like citizens' juries • Greater number of participants • Often used by governments to gauge opinions	• Conflicts resolved • Decision by majority vote • Eventually new policy options developed
Stakeholder decision analysis	• Combines quantitative multi-criteria and qualitative techniques	• Record and analysis of stakeholder's main concerns
Deliberative polling	• Public briefed on issue • Interview	• Survey of public's main concerns

facts. The "experts" often need training in how to communicate a complex subject to lay people. Among the important features of communication in such circumstances are the need to separate "need to know" from "nice to know" aspects, and how to deal with the emotive words and terms that have been used in the polarized discussion (Box 5.6). These approaches are directed to diverse audiences, taking into account factors such as age, culture, religion and education.

Communication with decision makers is usually one-to-one or in small groups such as advisory panels. Face-to-face communication to the public is often by addressing interest groups (e.g., farmers, consumers), community groups and schools in larger groups. Some countries have chosen to do this on a relatively formal basis to enhance the public understanding of science. Many of these formal discussions are held on a one-off basis and only involve a small part of the population.

BOX 5.6

EMOTIVE WORDS IN DISCUSSIONS ABOUT GM

Contamination/genetic pollution: This is a term widely used to indicate the mixing of transgenes with native genes as a result of gene flow from a GM crop, and/or post-harvest mixing of seed. It ignores the fact that these processes often occur between organic and non-GM crops and within GM crops.

Frankenfoods: This is a term often used in campaigns against GM crops to describe food products from those crops.

Genes: Some surveys have shown that many people think that GM tomatoes contain genes whereas non-GM tomatoes do not (see http://www.foodpolicyinstitute.org/docs/pubs/2004_Americans%20 and %20GM%20Food_Knowledge%20Opinion%20&%20Interest%20in%202004.pdf). This can suggest that some people consider genes to be "bad".

Genetic interfering: Suggests that the process of GM is unnatural. This should be compared with the use of mutagenesis (see Box 1.3) in conventional plant breeding.

Risk: Some people do not realize the difference between hazard and risk and do not recognize that there is no such thing as no risk (Chapter 2, Section I.C).

Scientists disagree: See text below on basic scientific method.

Superweeds: Weeds that can tolerate a particular herbicide. It ignores the fact that superweeds occur in conventional agriculture and are susceptible to other herbicides.

Terminator technology: A technology proposed for manipulating plants to ensure that the seeds were infertile. This would prevent farmers from saving seed and thus avoid buying commercial seed for the next planting season. The technology has never been commercialized so far. The technology can be compared with conventionally bred F1 hybrids (see Box 1.5) which are widely grown as they give higher yield; the parental genes segregate in seed saved from F1 hybrids and thus their yield and other characters revert back to the parental form with a consequent loss of farm productivity.

Unnatural: See Section II.B.2. The term ignores the fact that other technologies used in conventional and organic agriculture such as mutagenesis are equally unnatural.

Zombi seeds: used for terminator terminology.

However, even these forms of communication are not devoid of problems. Much of what needs to be conveyed to the public is new and can involve incomprehensible scientific concepts and terminology. This can be a reflection of the lack of scientific education in a rapidly changing world. Thus, the problem for scientists is to describe the facts in a clear, unambiguous manner. Furthermore, there is a lack of understanding by much of the general public of the basic scientific method, which is based on the following:

- Empirical observation;
- Development of hypothesis to explain a range of phenomena;
- Shown to be reproducible by independent experiment;
- Challenging the hypothesis;
- Development of new knowledge and theory.

In experimental sciences, the scientific method is based on challenging the interpretation of observations and facts and usually cannot be described in "black and white". A frequent point that opponents of the technology make is that "scientists disagree on this matter" even though the general scientific consensus may be criticized by only one scientist.

Understanding the "method" on which science is based gives the non-expert the ability to distinguish between "science" and "pseudoscience". The latter is based on anecdotal information and untestable hypotheses. Although the objectivity of scientific observation and results interpretation can be challenged on the basis of cognitive, sampling or cultural bias, there are numerous methods that ensure that such biases are minimized or eliminated. Furthermore, the objectivity of scientific findings is enhanced by the peer-review process which is a fundamental element lacking in "pseudo-scientific" literature.

V. LABELLING AND TRACEABILITY

Consumer choice is a fundamental right in most societies. In this context, labelling of GM-derived products can be an important element in enhancing the transparency of, and trust in, regulatory systems. However, labelling is not devoid of problems that are analysed below and brought up again in the context of the WTO agreement on Technical Barriers to Trade (see Chapter 6).

A. Why Label?

Labelling of GM foods is not part of risk assessment, although it is sometimes misperceived as such. For this reason, some countries such as the USA which limit food labels to safety-related information explicitly do not include information on GM content on food labels. However, there are two possible reasons for labelling GM products. First, some governments (particularly in the EU) see it as important that consumers be able to choose between foods on the basis of the presence or absence of GM-derived constituents and, for consumers to make the choice, they should have information about the food they are buying or consuming. An example of this recognition is in the preface to the European Union's (EU) Regulation 1830/2003/EC on GM food and feed labelling which mentions that consumer's choice is a reason underlying labelling of GM foods. Second, labelling is among the information required by the Cartagena Protocol (Annex II, paragraph 1 and Annex III, paragraph k) for the transboundary movement of LMOs (see Chapter 6), which has been accepted and signed by most nations involved in GM products. Of course, this form of labelling is not directed at consumers but rather at the immediate recipients of the LMOs in the recipient country.

However, as noted in Box 5.2, many consumers regard a label as a warning rather than a source of information.

B. How to Label?

By 2007, more than 40 countries had adopted regulations for labelling GM food. However, there is no internationally harmonized protocol for the labelling of GM foods leading to a multiplicity of national approaches. The only common feature is the requirement to label products derived from GM crops that are not substantially equivalent to their conventional comparator. The first major dichotomy in the national regulations concerning GM labelling is that in some countries labelling is voluntary whereas in others it is mandatory (Table 5.4).

The guidelines for voluntary labelling give rules that define which foods are called GM or non-GM, leaving the companies to decide if they want to use such labels on their products. Thus, there is an incentive to label beneficial attributes and to make "no content" claims for detrimental attributes. Mandatory labelling requires food handlers (retailers, processors and even sometimes food producers or restaurants) to label products or ingredients that contain, or are derived from, GM materials.

In countries that require mandatory labeling, the regulations differ in how products are labelled. First, countries differ in the types of food that have to be labelled which can include:

- Particular food ingredients or all ingredients that include detectable transgenic material;
- Highly processed products derived from GM ingredients, even without detectable transgenic material, for example oils from transgenic canola, sugar from transgenic sugar beet or sugar cane;
- Animal feeds;
- Meat and animal products from stock that has been fed with GM feed;
- Additives and flavourings;
- Food sold by caterers and in restaurants;
- Unpackaged food.

An example of this is shown in the EU labelling regulations (Box 5.7).

TABLE 5.4   Examples of labelling regulations.

Country	Labelling scheme	% threshold for unintended GM content	Are some biotech foods and process(es) exempt?
Canada	Voluntary	5%	N/A
USA	Voluntary	N/A	N/A
Argentina	Voluntary	N/A	N/A
Australia and New Zealand	Mandatory	1%	Yes, labelling only for detectable GM
EU	Mandatory	0.9%	Yes
Japan	Mandatory	5%[a]	Yes, labelling only for detectable GM
South Korea	Mandatory	3%[b]	Yes
Indonesia	Mandatory	5%	Yes

N/A, Not applicable;
[a]Relates to the top three ingredients;
[b]Relates to the top five ingredients.

BOX 5.7

GM LABELLING IN THE EU

In the EU, if a food contains or consists of genetically modified organisms (GMOs), or contains ingredients produced from GMOs, this must be indicated on the label. For GM products sold "loose", information must be displayed immediately next to the food to indicate that it is GM.

The GM Food and Feed Regulation (EC) No. 1829/2003 lays down rules to cover all GM food and animal feed, regardless of the presence of any GM material in the final product. This means that products such as flour, oils and glucose syrups have to be labelled as GM if they are from a GM source. Products produced with GM technology (cheese produced with GM enzymes, for example) do not have to be labelled. Products such as meat, milk and eggs from animals fed on GM animal feed also do not need to be labelled. Details on the labelling rules can be found in Table 1.

Any intentional use of GM ingredients at any level must be labelled. However, the Food and Feed Regulation provides for a threshold for the adventitious, or accidental, presence of GM material in non-GM food or feed sources. This threshold is set at 0.9% and only applies to GMOs that have an EU authorization. A temporary threshold of 0.5% for the presence of GM material not yet authorized but that has a favourable assessment

(cont'd)

BOX 5.7   (*cont'd*)

from an EU scientific committee expired in April 2007. This means that such unauthorized GM material cannot be present at any level.

Food or feed produced by a fermentation process using a genetically modified microbe (GMM) that is kept under contained conditions and is not present in the final product does not fall within the scope of this regulation. Such food and feed is considered to have been produced with the GMM, rather than from the GMM.

TABLE 1   Examples of labelling requirements under EC Regulation No. 1829/2003 for authorized GMOs (updated April 2008).

GMO type	Hypothetical examples	Labelling required?
GM plant	Chicory	Yes
GM seed	Maize seeds	Yes
GM food	Maize, soybean, tomato	Yes
Food produced from GMOs	Maize flour, highly refined soya oil, glucose syrup from maize starch	Yes
Food from animals fed GM animal feed	Meat, milk, eggs	No
Food produced with help from a GM enzyme	Cheese, bakery products produced with the help of amylase	No
Food additive/flavouring produced from GMOs	Highly filtered lecithin extracted from GM soybeans used in chocolate	Yes
Feed additive produced from a GMO	Vitamin B2 (riboflavin)	No
GMM used as a food ingredient	Yeast extract	Yes
Alcoholic beverages which contain a GM ingredient	Wine with GM grapes	Yes
Products containing GM enzymes where the enzyme is acting as an additive or performing a technical function		Yes
GM feed	Maize	Yes
Feed produced from a GMO	Corn gluten feed, soybean meal	Yes
Food containing GM ingredients that are sold in catering establishments		Yes (under EC Regulation 1829/2003)

In some countries, such as Australia, New Zealand and Japan, the labelling regulations only apply to products in which the GM protein can be detected. Thus, they would not apply to flour, sugar or oil from a GM crop. In the EU, these products have to be labelled but meat from animals fed on GM feed does not have to be labelled. Similarly, food from GM products does not have to be labelled in restaurants in countries such as Australia, New Zealand and Japan but it does in the EU; many restaurants get around this by stating in their menus that their food is not from GM sources. Thus, there are major international inconsistencies which are currently being considered by various international organizations.

A second major difference among countries in their approach to labelling is in the threshold level for labelling GM ingredients (Table 5.4). The threshold level is for the adventitious or accidental presence of products for which there is national authorization; usually it is 0% for products that have not been authorized. In Japan and South Korea, the threshold levels apply only to the major constituents of food.

Labelling of GM products has cost implications. With voluntary labelling schemes, these are borne by the consumers who wish to avoid GM food. In the mandatory schemes, all producers and consumers bear the cost.

The international production and trade of food and feed commodities often results in GM and non-GM product streams overlapping. There can be gene flow from a GM crop to a non-GM crop (see Box 4.6), and mixing in harvesting, storage and transport despite attempts to segregate the crops. Having threshold levels for unintended GM materials allows for such adventitious mixing but requires reliable detection techniques. These detection techniques are usually DNA or protein based (see Chapter 3, section III.C) and, although very sensitive, are limited in their ability to detect very small quantities of GM material. There are two related issues in international trade of GM commodities: segregation and identity preservation. The acceptable levels of mixing non-GM products with GM products are usually higher in segregation than in identity preservation.

C. Different Sorts of Labelling

Besides the labelling of GM foods as such, there are also other labels that make reference to the genetically modified organism (GMO) content of the labelled food product. For example, organically produced foods can be labelled as "organic" and generally must be free of GM ingredients, as in the case of EU legislation (Directive 2092/91/EC). Some member states of the EU, such as the Netherlands and Germany, have legal provisions in place that allow for products to be labelled as being "free of GMOs", although the EU policy has been to avoid such "negative labelling". The requirements of GMO-free labelling are similar to those for organic agriculture, so the GMO-free status does not rely upon GMO detection but rather on the adherence to protocols. Although not stated in the regulations, the reason for introducing this opportunity may be for farmers that are in the process of switching from conventional to organic but that have not been certified as organic, which may take several years.

In addition to these specific labelling requirements or options, in the EU (as in many other countries) all foods must comply with general food labelling requirements, including the declaration of ingredients, allergens, etc.

D. Traceability

The EU considers that the need to keep GM and non-GM food product streams separated and to label certain tolerances of adventitious mixing requires an audit trail so that a GM product can be traced from the farm to the consumer. The start of the product stream is the field in which the crop is grown. As discussed in Chapter 4, pollen can flow between crops and thus it is important to consider the coexistence of such crops to prevent "contamination" of a non-GM crop with pollen from a GM crop (Box 5.8).

The EU Directive 1830/2003 mandates that each stakeholder that produces or trades in GM raw materials, ingredients or foods is obliged to

---

BOX 5.8

COEXISTENCE

Coexistence refers to the ability to grow and manage along the supply chain both GM and non-GM crops in a way that avoids unwanted mixing and delivers products below predetermined market thresholds of "contaminants". It is not a safety issue. The three farming systems are involved as shown in Fig. 1.

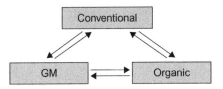

FIGURE 1   The three basic agricultural systems with arrows showing potential for gene flow among them and post-harvest mixing of their products.

The major issues include:

- Choice for farmers and consumers among the three basic farming systems;

(cont'd)

BOX 5.8   (*cont'd*)

- Mixing in the supply chain:
  - Pollen dispersal;
  - Seed mixing;
  - Volunteers and dropped seed;
  - Post-harvest mixing;
- Is it necessary for all agricultural products?
  - Non-foods, for example cotton;
  - Animal feed;
  - GM plants producing pharmaceutical and industrial products (which should be separated from food crops);
- Organic agriculture:
  - Organic agriculture is a self-contained system;
  - Many countries including those in the EU state that the presence of GMOs is not allowed in organic produce (but do not define limits);
  - UK Soil Association limit of 0.1% GMO;
- International trade:
  - Acceptability of GM produce in certain countries or regions;
  - World Trade Organization regulations (see Chapter 6).

Among the considerations for solutions are:

- Offer freedom of choice to those involved in the supply chain from farmers to consumers;
- Develop a system based on science and practical process management:
  - Gene flow information to define segregation distances;
  - Use of border and barrier rows;
  - Use of non-synchronous flowering of crop varieties;
  - How to deal with volunteers and contaminated farm machinery;
  - Post-harvest segregation;
- The system should be transparent:
  - Consult all stakeholders;
  - Possibility of a public register of GM crops;
- The system should minimize the impact on other farmers and consumers;
- The system should be on a case-by-case basis;
- There should be mechanisms for monitoring and reviewing the system:
  - Decide on allowable limits and detection systems;
  - Cost – who pays?
  - Have an initial period of intensive monitoring to gain experience.

pass information onto subsequent stakeholders in the food supply chain. Documentation is to be kept for 5–8 years. These traceability requirements for GMOs should facilitate both the withdrawal of products where unforeseen adverse effects on human health, animal health or the environment are established, and the targeting of monitoring to examine potential effects on, in particular, the environment. Traceability should also facilitate the implementation of risk management measures in accordance with the precautionary principle. Furthermore, the requirements would facilitate accurate labelling of GM food and feed products thereby ensuring that such information is available to operators and consumers and enabling control and verification of labelling claims. Requirements for labelling food and feed produced from GMOs should be similar in order to avoid confusion in cases of change in end use.

VI. LIABILITY AND REDRESS

As noted in section II.B, consideration has to be taken of the results of legal issues that may arise from the release of a GMO to the environment. Most cases (e.g., intellectual property rights and agreements between seed producers and farmers) can be dealt with using existing legislation in most countries. However, in two areas, there may be new legal issues: (1) the impacts of unintended or unpredictable hazards, and (2) the movement of transgenes from GM to non-GM crops causing what is regarded as "contamination". The importance of this issue varies from country to country and among different agricultural systems. For instance, "contamination" of an organic crop would be important where it affects the livelihood of the organic crop farmer but such "contamination" may not be as significant in mixed crop subsistence farming. We do not intend to discuss the legal side of liability and redress in such situations but we highlight some of the points that should be considered where this issue is important (Box 5.9).

One prominent liability situation related to GM crops was the "StarLink" corn incident described in Box 3.7.

Potential legal issues relating to different international agreements are discussed in Chapter 6.

VII. CONCLUSIONS

The importance of public perception in determining the acceptance of GM technology cannot be overestimated. While there has been little or no opposition to the use of modern biotechnology in medicine, the opposite is true when it comes to its uses in agriculture and the food industries.

BOX 5.9

_____

SOME ISSUES RELATING TO LIABILITY AND REDRESS

Among the points that need to be considered are:

- Who is entitled to claim redress;
- What can be claimed:
  - Loss in crop value;
  - Non-compliance with coexistence scheme;
  - Punitive damages;
- Who pays
  - Redress under existing laws (farmer or supplier);
  - Voluntary industry-led scheme;
  - Statutory redress scheme.

One of the reasons for this is that in the case of the latter there have been few obvious consumer benefits. The "first generation" of agricultural biotech products has so far primarily benefited the agricultural sector rather than consumers.

The continuing controversy over the "first generation" of agricultural biotechnology products is clouding the future of "second generation" products containing "output" traits, such as improved quality or enhanced nutrient content, which could be of major benefit to consumers worldwide (see Chapter 1). Although a number of such products are already at an advanced stage of development, public opposition threatens to slow down the pace of innovation. Unreasoned public hostility towards biotechnology may have an adverse impact both on the innovation process (i.e., products that are badly needed may never be developed) and on national regulation with concomitant effects on technology transfer and trade.

The ethics surrounding the application of modern biotechnology also constitute an important influence on public perception. The reason for this is that biotechnology, through its power to change the course of evolution or to speed it up, is often perceived as a challenge to basic ethical values and the notion of "naturalness". Some applications of biotechnology can only be legitimized through a constructive debate among stakeholders focusing on those issues that guide decision making and action while maintaining a central role for scientific information and analysis.

One of the main decisions that regulators have to make is whether a perceived risk outweighs the benefits of using the technology. The identification and levels of risks come mainly from the analyses described in

Chapter 2 and would not vary much from country to country (except, for example, on some ecological aspects such as releases in centres of origin or diversity of a crop species). However, the benefits vary significantly from country to country depending on factors such as food security, poverty and impact on potential for international trade. In the overall risk/ benefit analysis, the final decision on the release of a GM crop was summarized by the Nuffield Council on Bioethics (2006) as: "The question is not what if; it is what if not."

References

Aerni, P. and Bernauer, T. (2006). Stakeholder attitudes toward GMOs in the Philippines, Mexico and South Africa: the issue of public trust. *World Development* **34**(3), 557–575.

Bruce, D. and Bruce, A. (1998). *Engineering Genes: The Ethics of Genetic Engineering*. Earthscan Publications Ltd, London.

CEC (2004). Maize biodiversity: the effects of transgenic maize in Mexico. Secretariat Article 13 report. Commission for Economic Cooperation of North America.

Cormik KL, C. (2003). Perceptions of risks relating to biotechnology in Australia. *Int. J. Biotech.* **5**, 95–104.

Economic and Social Research Council (1999). The politics of GM food. Risk, science and public trust. Special Briefing No. 5.

Ekanem, E., Muhammed, S., Tegegne, F. and Singh, S. (2004). Consumer biotechnology food and nutrition information sources: the trust factor. *J. Food Distrib. Res.* **35**, 71–77.

European Federation of Biotechnology (2004). Who should communicate with the public and how? http://files.efbpublic.org/downloads/EU_FW_REPORT.pdf

Frewer, L.J., Howard, C., Hedderley, D. and Shepherd, R. (1996). What determines trust in information about food-related risks? Underlying psychological constructs. *Risk Analysis* **16**, 473–486.

Gaskell, G., Allum, N. and Stares, S. (2003). Europeans and Biotechnology in 2002. Eurobarometer 58.0. European Commission, Brussels.

Gaskell, G., Stares, S., Allansdottir, A., Allum, N., Corchero, C., Fischler, C., Hampel, J., Jackson, J., Kronberger, N., Mejlgaard, N., Revuelta, G., Schreiner, C., Torgersen, H. and Wagner, W. (2006). Europeans and Biotechnology in 2005: Patterns and Trends, Final report on Eurobarometer 64.3. A report to the European Commission's Directorate-General for Research, Brussels. http://www.ec.europa.eu/research/press/2006/pr1906en.cfm

Hoban, T. (1998). Trends in consumer attitudes about agricultural biotechnology. *AgBioForum* **1**(1), 3–7. Available on the World Wide Web: http://www.agbioforum.org.

Hoban, T. (2004). Public Attitudes towards Agricultural Biotechnology. ESA Working Paper No. 04-09. http://www.fao.org/docrep/007/ae064e/ae064e00.htm

Hunt, L.M., Fairweather, J.R. and Coyle, F.J. (2003). Public understanding of biotechnology in New Zealand: factors affecting acceptability rankings of five selected biotechnologies. Research Report No. 266. Lincoln University, New Zealand. http://hdl.handle.net/10182/738

IFIC (2008). International Food Information Council: Food Biotechnology. A study on US consumer trends. http://www.ific.org/research/biotechres.cfm

Madsen, K.H., Holm, P.B., Lassen, J. and Sandøe, P. (2002). Ranking genetically modified plants according to familiarity. *J. Agric. Environ. Ethics* **15**, 267–278.

Nuffield Council on Bioethics (2006). The use of genetically modified crops in developing countries: a follow-up. www.nuffieldbioethics.org

Pidgeon, N., Kasperson, R.E. and Slovic, P. (2003). *The Social Amplification of Risk*. Cambridge University Press, Cambridge, UK.

Suwannaporn, P., Linnemann, A. and Chaveesuk, R. (2008). Consumer preference mapping for rice product concepts. *Brit. Food J.* **110**, 595–606.

UK Cabinet Office. (2008). Food matters: towards a strategy for the 21st century. UK Cabinet Office, July 2008.

Further Reading

AEBC (2003). GM crops: co-existence and liability. A report by the Agriculture and Environment Biotechnology Committee. http://www.aebc.gov.uk/aebc/reports/coexistence_liability.shtml

Brossard, D., Shanahan, J. and Nesbitt, T.C. (eds) (2007). *The Public, the Media and Agricultural Biotechnology.* CABI, Oxford, UK.

Demont, M. and Devos, Y. (2008). Regulating coexistence of GM and non-GM crops without jeopardizing economic incentives. *Trends Biotechnol.* **26**, 353–358.

Gardner, D. (2008). *Risk: The Science and Politics of Fear.* Virgin Books Ltd, London.

Gaskell, G., Bauer, M.W., Durant, J. and Allum, N.C. (1999). Worlds apart? The reception of genetically modified foods in Europe and the US. *Science* **285**, 384–387.

Hails, R. and Kinderlerer, J. (2003). The GM public debate: context and communication strategies. *Nature Rev.: Gen.* **4**, 819–825.

Lemaux, P.G. (2008). Genetically engineered plant and foods: a scientist's analysis of the issues (part 1). *Annu. Rev. Plant Biol.* **59**, 771–812.

McHugen, A. (2007). Public perceptions of biotechnology. *Biotechnol. J.* **2**, 1105–1111.

Slovic, P., Finucane, M.L., Peters, E. and MacGregor, D.G. (2004). Risk as analysis and risk as feelings: some thoughts about affect, reason, risk, and rationality. *Risk Analysis* **24**, 311–322.

ABSTRACT

The biosafety of GMOs is controlled by a range of interlinked policies and legal, administrative and technical instruments. This chapter describes how such regulations are drawn up and implemented and how they interact with other international agreements.

OUTLINE

I. INTRODUCTION

A. National Biosafety Frameworks (NBFs) and Constituent Elements

The smooth implementation of biotechnology regulations depends on a number of interlinked policies and legal, administrative and technical instruments that constitute what is commonly referred to as a National Biosafety Framework (NBF). Five distinct elements are an essential part of an NBF as summarized below.

1. Biosafety Policy

Biosafety policies are normally part of a broader set of policies related to biotechnology, industrial development, agriculture, trade, health and environmental protection. The fact that these broader policies have been created at different times, with different objectives in mind and are administered by different bodies, makes it imperative to establish national biosafety policies which lie at the interface of these other policies, thereby ensuring harmonization and the functionality of implementation mechanisms.

2. Legal Instruments

Acts or Decrees, and the complementary technical regulations and guidelines, provide the necessary legal base and authority to implement regulatory oversight. Such authority may be given by existing regulatory regimes or amending existing regimes, or by the promulgation of GMO-specific regulations *de novo* (Table 6.1).

In addition, regulatory systems establish the boundaries of regulatory oversight and enforcement; that is, they define the regulated article (product, process and application) and provide the background for the development of technical instruments such as guidelines for risk assessment and the issuing of permits. With regard to risk assessment, regulations define the breadth of the assessment; that is, whether the assessment is conducted on purely a scientific basis or whether it also includes risk/benefit analysis and consideration of ethical and socio-economic issues.

Last but not least, regulatory systems must be consistent with and in compliance with international agreements and norms (e.g., WTO, Cartagena Protocol, *Codex Alimentarius*, etc.) (see Section III.A).

3. Administrative System

An administrative infrastructure is necessary to support the implementation of regulatory systems which, in turn, involve the establishment of mechanisms for risk analysis, risk management, post-commercialization

TABLE 6.1   Approaches to setting up a National Biosafety Framework.

Regulatory options	Advantages	Disadvantages
Existing regulations	Potentially allows the regulatory scrutiny of products with novel traits developed through conventional technologies (see also Section I.C.1 on "product-based regulation").	The scope of existing legislation may be inadequate to deal with GM products OR Problems of coordination may arise from the involvement of two or more authorities with overlapping jurisdictions.
De novo **system**	Streamlined specifically for GM products, thus addressing public concerns with regards to GMOs and derivative products.	Novel traits derived by means of conventional technology escape regulatory scrutiny on account of the narrow focus of regulation (see also section I.C.2) OR Products for which there is extensive familiarity are subject to disproportionate regulatory scrutiny.

monitoring and risk communication, as well as mechanisms to handle notifications or requests for authorization for activities pertaining to the development, use and commercialization of GMOs and derivative products.

The functionality of administrative systems depends on:

• The existence of guidelines that make the different components of the regulatory system operational;
• Access to up-to-date scientific information and expertise for risk assessment;
• Feedback mechanisms ensuring that the system responds to changing circumstances (e.g., scientific developments and public attitudes).

NBFs are essential in defining the administrative infrastructure entrusted with regulatory implementation. The remits and functions of committees that are established in the administrative infrastructure are reviewed in detail in Section II.B.

4. Monitoring Systems and Enforcement

Post-commercialization monitoring and general surveillance are mechanisms to deal with uncertainties regarding the long-term impacts, including benefits, arising from large-scale release of GMOs. Monitoring systems define the requirements for post-approval review and, if necessary, additional information for risk assessment and the conduits through which monitoring results are communicated to relevant authorities, experts and the public. These subjects have been reviewed extensively in Chapter 2, and specifically for human and animal health and for environmental release assessments in Chapters 3 and 4, respectively. Enforcement mechanisms are also necessary to determine levels of inspections and audit, as well as for the implementation of measures to impose administrative, monetary or trade penalties.

5. *Public Involvement*

The need for public participation in, and for access to, information related to environmental issues is highlighted in Article 10 of the Rio Declaration on Environment and Development (http://www.unep. org/Documents.Multilingual/Default.asp?DocumentID=78&ArticleI D=1163). The communication of regulatory procedures and decisions to the public contributes to increased public awareness and perception of the technology, and contributes to the transparency and legitimacy of the institutions involved. The role of public perception in shaping regulations and decisions regarding GMOs and GM products is analysed in detail in Chapter 5.

However, the practicalities of public involvement are not trivial and have a bearing on operational mechanisms and procedures. The latter have to address the following questions:

• How are public inputs solicited and how are comments and responses to comments recorded?
• How are public inputs reflected in regulatory decisions?

Several regulatory systems such as that of the USA make provisions for information access through public registers and public participation in advisory committees and contributions in the risk assessment process. In designing participatory systems, care should be exercised to have balanced representation of different stakeholder groups, for example academia, industry and non-governmental organizations.

A comprehensive guide to establishing an NBF is given in UNEP-GEF (2005).

B. Evolution of NBFs

As described in Chapter 1, section VI, the need for regulation in biotechnology arose from the realization that recombinant technologies were capable of creating organisms with novel characteristics for which there was little or no experience regarding potential impacts on human health and the environment. Extensive debate occurred as to whether the technology itself and its products warranted new regulations specific to biotechnology, resulting initially in guidelines which were later given the force of law.

There are two major factors that have led to the development of national regulatory frameworks: national factors and international agreements. Public and scientific pressure in the USA led to the National Institutes of Health (NIH) issuing Guidelines on Recombinant DNA Research on 23 June 1976, which provided for both physical and biological containment protocols. Four levels of physical containment governed rDNA (recombinant DNA) laboratory experiments, requiring protective measures ranging from gloves to extractor hoods and, at the highest containment level, isolated rooms with separate ventilation and water systems, lower barometric pressure and air-locks. They also provided for three levels of biological containment, requiring that organisms were purposely modified so that they could not survive outside the laboratory. Experiments involving DNA from highly pathogenic bacteria or genes coding for toxins were prohibited outright. The NIH guidelines became an international standard of reference for researchers in academia and industry. Because they, and their subsequent revisions, were adopted only after lengthy public hearings, the guidelines also reflected unprecedented public input into scientific matters.

To provide a legal basis for these guidelines, the US government enacted various laws between 1976 and 1979. The principles behind these recombinant DNA laws were taken up by Canada and the EU and form the basis for many other national GM regulatory frameworks. The worldwide debate on the safety of GM technology led to a seminal document "Recombinant DNA Safety Considerations" (OECD, 1986) and its follow-up "Safety Considerations for Biotechnology" (OECD, 1992). These documents have since influenced the development of national biosafety regulations *ab initio* and the evolution of existing regulations to cover biotechnology and its products.

C. The Regulatory Trigger

1. Product-Based Regulations

Typical examples of regulatory evolution to encompass the products of recombinant technologies are the regulatory systems of Canada and the USA. They reflect the OECD recommendation that there is no need for countries to develop new regulations for biotechnology as "there is no scientific basis for specific legislation to regulate the use of recombinant DNA organisms".

In Canada, product-based Acts (e.g., for food, feeds, fertilizers, pesticides) that pre-existed the advent of recombinant biotechnology were adapted to cover biotechnology applications as the latter were seen as merely different approaches to produce new lines within a given family of products. As a consequence, regulatory oversight is triggered when a biotechnology-derived product is considered to be novel (termed plants with novel traits; PNTs) in the Canadian environment. The same applies for plants with newly introduced traits regardless of whether these have been introduced by recombinant or conventional technologies. Regulatory oversight is covered by the Novel Foods Regulation under the Canadian Foods and Drugs Act and is triggered whenever trait or product is considered to be novel in the Canadian environment (http://www.hc-sc.gc.ca/fn-an/legislation/acts-lois/fdr-rad/division-titre28-eng.php). An overview of the Novel Foods Regulation is given in Fig. 6.1.

The introduction of plants or micro-organisms with novel traits as a food source, and the importation of novel whole foods and food ingredients, require mandatory pre-market notification if they were previously not available in Canada or have been genetically modified from a pre-existing counterpart through a change that is considered to be major. It should be noted that the Canadian product-based regulation in biotechnology is more expansive than other product-based regulatory systems in that it also covers organisms that have been modified through non-recombinant technologies provided that the latter are considered to be novel traits or ingredients that constitute a major departure from the non-modified parent organisms. Here the crucial point is how "novelty" and "major change" are defined (see Box 6.1); see also Case Study 1, Appendix D.

In the USA, the NIH established an rDNA Advisory Committee (RAC) to assess the state of knowledge. On the basis of the recommendations of the RAC, a more relaxed set of research guidelines was published by NIH in 1983. These guidelines continue to be referenced by industry, federal and academic laboratories.

In 1983, the NIH approved the first environmental release of an organism developed using rDNA technology (ice-minus strain of the bacterium *Pseudomonas* to control freezing damage in strawberries). The release generated immense controversy for failing to prepare a statement or assessment of the environmental impact of NIH's regulatory decision as required under the National Environmental Policy Act (NEPA). Since then, all responsibility that NIH had for regulating environmental introductions of GMOs was relinquished despite the fact that it was unclear which federal regulatory agency or agencies would be responsible for such introductions.

In response to a need for clarification, the Office of Science and Technology Policy (OSTP) began a process culminating in 1986 in the "Coordinated Framework for Regulation in Biotechnology". According to this Framework, the products of rDNA technology should be regulated on the basis of the unique characteristics and features that they exhibit, not their method of production. The products of rDNA technology were considered to pose risks to human and environmental health similar to those posed by conventional products already regulated within the USA. The possibility to develop new guidelines, procedures and even

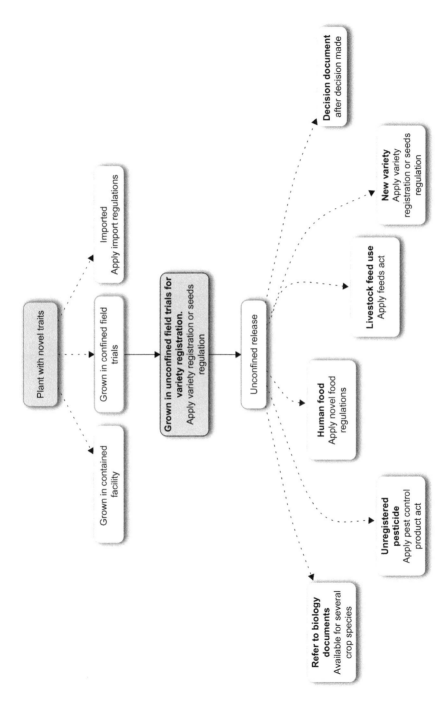

FIGURE 6.1 Diagram showing the interlinking of considerations in the Canadian Novel Foods Regulations (see colour section).

regulations to supplement or alter the scope of existing statutes was not ruled out. The Coordinated Framework identified three federal agencies as having primary responsibilities for evaluating the products of rDNA technology.

BOX 6.1

DEFINITION OF TERMS IN THE CANADIAN NOVEL FOOD REGULATIONS (HTTP://WWW.HC-SC.GC.CA/FN-AN/ LEGISLATION/ACTS-LOIS/FDR-RAD/ DIVISION-TITRE28-ENG.PHP)

"**genetically modify**" means to change the heritable traits of a plant, animal or microorganism by means of intentional manipulation (modifier génétiquement).

"**major change**" means, in respect of a food, a change in the food that, based on the manufacturer's experience or generally accepted nutritional or food science theory, places the modified food outside the accepted limits of natural variations for that food with regard to:

(a) The composition, structure or nutritional quality of the food or its generally recognized physiological effects;

(b) The manner in which the food is metabolized in the body; or

(c) The microbiological safety, the chemical safety or the safe use of the food (changement majeur).

"**novel food**" means:

(a) A substance, including a microorganism, that does not have a history of safe use as a food;

(b) A food that has been manufactured, prepared, preserved or packaged by a process that:
   (i) Has not been previously applied to that food, and
   (ii) Causes the food to undergo a major change; and

(c) A food that is derived from a plant, animal or microorganism that has been genetically modified such that:
   (i) The plant, animal or microorganism exhibits characteristics that were not previously observed in that plant, animal or microorganism,
   (ii) The plant, animal or microorganism no longer exhibits characteristics that were previously observed in that plant, animal or microorganism, or
   (iii) One or more characteristics of the plant, animal or microorganism no longer fall within the anticipated range for that plant, animal or microorganism (aliment nouveau).

In 1992, OSTP released another document entitled "Exercise of Federal Oversight within the Scope of Statutory Authority: Planned Introductions of Biotechnology Products into the Environment", outlining the proper basis by which federal regulatory agencies were expected to exercise their regulatory authority. According to this, if more than one federal regulatory agency has an interest in a particular product, lead agencies are identified as being responsible for coordinating activities to limit any potential duplication of effort.

The jurisdiction of the different agencies is shown in Table 6.2.

The jurisdiction of the regulatory agencies is determined by the regulatory mandate of the respective agencies, as well as the intended use of the GMO. Consequently, the safety review process may or may not

TABLE 6.2 Overview of responsible agencies under the coordinated framework.

Responsible agency	Jurisdiction	Regulatory trigger
Food and Drug Administration (FDA)	Food and food additives; feed and veterinary drugs	• Intentional and unintentional adulteration of food and food components with substances considered poisonous or hazardous to human health. A food or food component is considered adulterated if reasonable certainty exists that its consumption may have deleterious effects on human health. • Substances intentionally added to foods that are not generally recognized as safe (GRAS) based on prior scientific testing or historical use, or that are not otherwise exempt (e.g., pesticides, etc.). A substance may be considered as a food additive if determined to be significantly different in structure, function or amount from a substance already consumed as part of the diet or lacks a sufficient history of safe use.
US Department of Agriculture. Animal and Plant Health Inspection Agency (USDA/APHIS)	Plant pests, plants, veterinary products	• For a transgenic plant to be considered as a regulated article, any of the donor or recipient organism, vector or vector components must be in the list of plant pests or noxious weeds regulated under the Federal Noxious Weeds Act.
US Environmental Protection Agency (US EPA)	Planting and food/feed uses of pesticidal plants; new uses of existing pesticides, novel micro-organisms	• Pesticidal substances intended to be produced and used in living plants, or in plant-derived products, and the genetic material necessary for the production of such a pesticidal substance. • Genetically modified microbial pesticides, i.e. bacteria, fungi, viruses, protozoa, or algae, whose DNA has been modified to express pesticidal properties. The modified micro-organism generally performs as a pesticide's active ingredient.

involve all three agencies. Finally, it should be noted that the distinction between product- and process-based regulation is blurred in the case of transgenic plants developed through transformation technologies using *Agrobacterium* as the transformation vector. The latter is included in the list of plant pests and, therefore, any such product is regulated by USDA/APHIS.

2. Process-Based Regulations

In contrast to product-based regulations, process-based ones adopt a philosophy that existing legislation is not sufficient to cover products and applications arising from the use of rDNA technologies. An example of process-based regulations is that of the European Union, the operational framework of which is principally defined by Directive 2001/18/EC and a number of specific regulations. In addition, the ratification of the Cartagena Protocol effectively means the adoption of process-based regulations by over 150 countries (see below). The Protocol deals with the transboundary movement of living modified organisms (LMOs) which are defined as "any living organism that possesses a novel combination of genetic material, obtained through the use of modern biotechnology". It defines "modern biotechnology" as the "**a**. *in vitro* nucleic acid techniques, including recombinant DNA and direct injection of nucleic acid into cells or organelles, or **b**. fusion of cells beyond the taxonomic family...".

A number of process-based systems adopt the Precautionary Principle (see Box 2.3) as a guide. For example, the EU Directive 2001/18/EC explicitly adopts the Precautionary Principle and requires the evaluation of long-term and indirect effects, as well as impacts arising from changes in agricultural practice. Implicitly, however, it recognizes that the Precautionary Principle may be difficult to apply and, counterbalancing this, evokes the general principles of risk management such as proportionality, non-discrimination, consistency, and costs and benefits arising from actions or inactions. In addition, some countries (e.g., Norway, New Zealand and the EU (Commission of the European Communities, 2000)) contain options or requirements for balancing or mitigating risks associated with GM crops with potential environmental benefits arising from their cultivation. For example, GM crops with pesticidal traits may be regarded as preferable to conventional pest management with the use of chemicals with respect to impacts on non-target organisms. On the other hand, changes in agricultural practice associated with the use of herbicide-tolerant GM crops may result in more efficient weed control and, potentially, reductions in biodiversity. Consequently, this type of risk–benefit analysis must be conducted on a case-by-case basis comparing GM crop use and management with conventional agricultural practices taking into account regional differences and farming systems. This type of risk–benefit

TABLE 6.3 Comparison of process-based and product-based regulations.

Regulatory trigger	Assumptions	Advantages	Disadvantages
Process based	GM technology represents new sets of risks	Single authority offers a better coordination mechanism	Conventional products that are potentially risky can escape the net of regulation; regulation lags scientific progress leading, potentially, to overregulation.
Product based	"there is no scientific basis for specific legislation to regulate the use of recombinant DNA organisms"	Focus on phenotypic characteristics	Choice of comparator difficult and as such establishing familiarity and/ or substantial equivalence difficult (see Chapter 2).

analysis requires the simultaneous operation of environmental risk monitoring and risk–benefit systems.

Socio-economic risk assessment is required in some countries (e.g. Norway (Rosendal, 2008) and the Philippines (Ochave and Estacio, 2001)) but how this is to be conducted is not always clear. Furthermore, socio-economic risk assessment in the case of importation of GM commodities would contravene the provisions of the WTO Agreement on the Application of Sanitary and Phytosanitary Measures (see below).

Advantages and disadvantages can be identified in both systems (see Table 6.3). In any case, even though the underlying philosophies of product- and process-based regulations are fundamentally different, the information requirements for risk assessment are similar and may differ only in the degree of detail, particularly in the requirements for molecular characterization (see Section II.C).

Therefore, there is a growing trend of regulatory harmonization which is achieved through international agreements and negotiations (see below) and standards-setting bodies. For example, in food safety assessment, the Canadian, EU and US systems accept as their foundation principles developed by *Codex Alimentarius* such as the comparative assessment of the GMO with its best conventional counterpart that has a known history of safe use (Paoletti *et al.*, 2008).

II. IMPLEMENTATION OF NBFS

A. Implementation Overview

NBFs set out the system by which GMOs have to be approved on safety grounds and, to this end, each GMO is subjected to science-based risk analysis. This ensures that the release and marketing of GMOs only takes place with the explicit consent of regulatory authorities.

Regardless of the regulatory trigger, risk assessment strategies are based on a common set of principles and guidelines which are described in detail in OECD (1993) and *Codex Alimentarius* (2003), and are extensively reviewed by Paoletti *et al.* (2008).

The basic guidelines are:

- Triggers are needed to start a risk assessment;
- The assessment should follow a structured and integrated approach;
- New hazards of the GMO when compared with a conventional counterpart should be identified;
- Both intended and unintended effects of the GMO when compared with a conventional counterpart should be evaluated.

Although the implementation of NBFs varies among countries, there are a number of common elements which are summarized in Fig. 6.2.

B. The role of National Biosafety Committees

Many countries have biosafety structures that incorporate control and oversight at both the local and national levels, usually comprised of local (company or institutional) biosafety committees and a National Biosafety Committee (NBC). Some countries also have regional (state or province) biosafety committees which act between the local and national committees.

The remit of the local biosafety committee is usually to advise workers in the company or institute on biosafety matters and to oversee local adherence to biosafety regulations and conditions of controlled and contained releases. Regional biosafety committees act as a link between the local and national committees and also cover regional biosafety matters.

The NBC reviews, and conducts risk assessments on, applications for GMO releases and advises the decision makers on the biosafety issues in relation to the application. The legislation setting up most NBCs usually lays out their terms of reference, the review procedure and the mode of operation.

The terms of reference of NBCs are usually defined in national legislation, regulations or guidelines, and specify:

- The nature of the NBC, that is whether it functions as subordinate to an executive body (e.g., Ministry, Agency, etc.) and has advisory functions, or is independent having executive functions;
- The areas of competence of the NBC, for example field testing, production, importation, export and commercialization of GMOs and

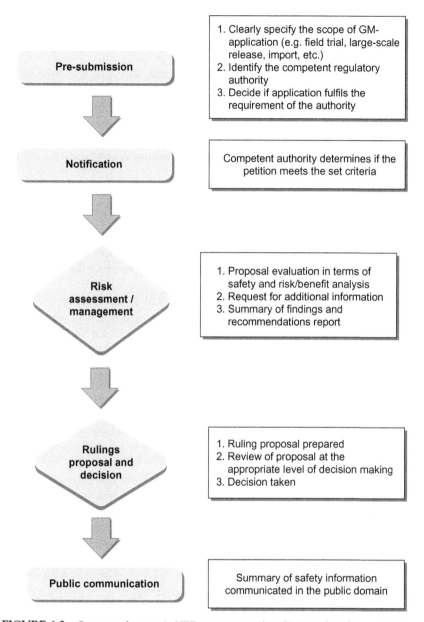

FIGURE 6.2   Common elements in NBFs assessment of applications for release of GMOs.

their derivative products. Different subcommittees may deal with each of these aspects;
• Whether reviews involve only the scientific evaluation of risks or are broader, including benefits assessment.

The review procedure usually comprises:

- Description of the context in which the review is conducted;
- Scientific evaluation of potential risks based on literature, and experimental and field trial data submitted by the applicant. Risk–benefit analyses should be subject to the terms of reference of the NBC;
- Communication of the consequences of potential risks in a decision document;
- Establishment of post-commercialization monitoring plans (if appropriate);
- Specification of the conditions under which an approval should proceed.

The mode of operation lays out:

- Rules of procedure, information management and documentation, and how conflicts of interest and confidential business information are dealt with;
- Membership of the NBC and nature of requisite expertise, for example what sort of scientific experts are needed (life and analytical sciences, ecology, agronomy), other necessary expertise (legal experts, representatives of executive bodies), and whether members of the public should be included.

C. Applying for a Release Permit

The basic regulatory framework sets out the procedures to apply for a GMO release, for the assessment of that release and for the decision making on that release.

The scope of the application will depend upon the purpose of the release. In general, for contained releases the focus will be on environmental biosafety as there will be little need for information on food and feed safety (except if there is a possibility of the GMO entering the food/feed chain). For commercial release, information will be required on both food/feed and environmental safety.

Usually, the applicant has to provide the following information on the GM product:

- Description of the recipient plant; biology of the recipient plant, uses of the recipient plant as food and/or feed, agronomy of the recipient plant;
- Description of the donor organism, for example is it a pathogen? Does it produce a toxin or allergen?
- Description of the genetic modification, including a molecular description of the construct(s), vector(s) used; gene(s), promoter(s),

terminator(s), selection marker(s), methods to determine its purity; transformation procedure;

- Description of the product, including copy number of insert, determination of site(s) of insertion; expression of RNA; expression of protein if gene inserted;
- Safety assessment of product for food and feed safety, including toxicological and allergenicity tests as described in Chapter 3; compositional analysis of key components; evaluation of metabolites; nutritional modifications; possible effects of food processing;
- Safety assessment for environmental safety as described in Chapter 4; possible gene flow to wild species in area of release, possibility of weediness, possible impact on non-target organisms;
- Description of the scope of the release; if it is a contained release, the information should include the purpose of release, site and size of release, adjacent agricultural crops, measures for containing the release (e.g., fencing, control of access), disposal of GM material, measures for cleaning up site after release finished, monitoring measures both during release and for a set period after the release, proposed actions in case of emergency. If it is a commercial release the information should include the purpose of the release, stewardship agreements between company and growers, monitoring measures.

A more detailed description of a typical guidance document for the information required to make a risk assessment is in Appendices B and C.

D. Implementation Constraints

Regulatory oversight for a single GM product lies within the area of competence of several different government authorities/agencies (e.g., Ministries of Agriculture, Environment, Health, Trade, and, on occasions, Science and Technology, etc.). More often than not, this creates problems of coordination among the different authorities, the consequences of which are delays in product approvals and escalation of regulatory costs. A recent study (Kalaitzandonakes *et al.*, 2007) estimates the costs of regulatory compliance for insect-resistant corn and herbicide-tolerant corn to lie in the range of US$6–15 million, and costs related to the molecular characterization of the modified event and of stewardship plans appear to be escalating over the years. In another study (Pray *et al.*, 2005), which may be more representative of the situation in developing countries with a domestic seed industry and the capacity to develop new biotechnology crops, the authors report the costs of regulatory compliance in India. The costs incurred range from US$2 to 4 million in the case of private firms and US$50 to 60 thousand in the case of government research institutions. This represents a cost reduction of almost two orders of magnitude for the public sector, which is attributable to the fact that salaries have

not been factored in and biosafety testing conducted by national institutes is done for a nominal cost. Furthermore, the years of delay to obtain
release permits can be a concern. This represents a major disincentive for
product development and, as such, an indirect cost of regulation. It may
be one of the reasons why very few crops of direct relevance to developing world needs have been commercialized to date. Direct and indirect
costs of regulation may become prohibitive and result in non-competitive
market structures for biotechnology-derived products.

However, there are differences among developing countries. Broadly
speaking, one can identify three tiers: those countries that have capacity to develop new technologies (see Case Study 3, Appendix D; those
that have capacity to adopt technologies developed elsewhere; and those
with minimal or no capacity to adopt new agricultural technologies.
Inevitably, resources, infrastructure and policies differ widely among
these three groups of countries, as do priorities in setting up and implementing regulations for the activities described above. Even in the more
advanced developing countries, there is a lack of expertise and infrastructure to conduct science-based risk assessment. Additionally, deficits
in financial resources and expertise are compounded by the lack of institutional transparency that is necessary to legitimize decisions.

E. Cooperation in GMO Regulatory Oversight Between Countries

The problem of inadequate human and financial resources to conduct
science-based risk assessment could, in theory, be overcome by pooling
of expertise and harmonization of risk assessment procedures at the
regional and subregional level. However, in practice this has not proven
possible, possibly because such an approach is seen by countries as ced-
ing their sovereignty in taking regulatory decisions or due to differing
policies regarding GMOs and/or a variety of administrative obstacles.
The issue of regulatory sovereignty is debatable as obligations to international trade agreements have to be taken into account. Nevertheless,
significant steps promoting international cooperation have been or are
being achieved through the Advance Informed Agreement and Biosafety
Clearing House Mechanisms of the Cartagena Protocol and international
standards-setting bodies, such as the *Codex Alimentarius* (see Section III
below).

F. Grey Areas of Regulation

Although risk assessment frameworks for GMOs are broadly similar
across countries and regions, and considerable effort goes into harmonizing elements of the risk assessment where appropriate, important differences do exist among countries. Some of these differences relate back to

the distinction between regulating on the basis of process or product discussed in Section I.C, while others are more a function of the pre-existing regulatory systems that have been adapted for use with GMOs. This is not the place to describe these differences in detail but it is important to be aware that there are such grey areas of regulation.

For example, as discussed in Chapter 4, GMOs with multiple traits, known as stacked products, may be viewed in a number of ways. In particular, where multiple traits are combined through conventional breeding, regulatory systems in some countries such as Canada and the USA view the resulting product as a simple combination of the individual traits that may require little additional regulatory assessment beyond what is needed for the individual traits. The EU and some other countries regard the stacked trait combination as an entirely new product requiring a full separate assessment.

Obviously, these different approaches have significant consequences for the ability to gain regulatory approvals of stacked trait products in different countries. These sorts of differences in regulatory systems also create challenges for international initiatives aimed at harmonizing regulatory approaches.

III. BEYOND NATIONAL REGULATIONS: INTERNATIONAL INSTRUMENTS OF BIOTECHNOLOGY REGULATION

A. Multilateral Agreements

1. The Convention on Biological Diversity

The Convention on Biological Diversity (CBD) (http://www.cbd.int/convention/guide.shtml) is an international treaty, the main objectives of which are: "the conservation of biological diversity, the sustainable use of its components, and the fair and equitable sharing of the benefits from the use of genetic resources". The Convention was adopted in 1992 and has been signed and/or ratified by over 190 Parties (countries).

The Convention deals with biotechnology in two articles (Box 6.2).

The Conference of the Parties, in its 2nd meeting in Jakarta, Indonesia, in 1995, initiated negotiations to establish the protocol set out in Article 19 which resulted in the adoption of the Biosafety Protocol, also known as the Cartagena Protocol, in January 2000. Currently, there are over 150 Parties to the Protocol (see http://www.cbd.int/biosafety/signinglist.shtml).

2. The Cartagena Protocol

The Protocol focuses specifically on transboundary movements of living modified organisms (LMOs) (defined as non-processed GMOs that are viable if introduced in the environment, for example seed and

---

BOX 6.2

ARTICLES IN THE CONVENTION ON BIODIVERSITY RELEVANT TO BIOTECHNOLOGY

Article 8(g) contains an obligation of the Parties to "Establish or maintain means to regulate, manage or control the risks associated with the use and release of living modified organisms resulting from biotechnology which are likely to have adverse environmental impacts that could affect the conservation and sustainable use of biological diversity, taking also into account the risks to human health".

The first paragraph of Article 19 addresses the need for the proper handling of biotechnology and the distribution of its benefits by taking appropriate measures for "the effective participation in biotechnological research activities by those Contracting Parties, especially developing countries, which provide the genetic resources for such research, and where feasible in such Contracting Parties".

The second paragraph of Article 19 requires that the Contracting Parties "take all practicable measures to promote and advance priority access on a fair and equitable basis by Contracting Parties, especially developing countries, to the results and benefits arising from biotechnologies based upon genetic resources provided by those Contracting Parties. Such access shall be on mutually agreed terms".

The third paragraph of Article 19 requires the Parties to "consider the need for and modalities of a protocol setting out appropriate procedures, including, in particular, advance informed agreement, in the field of the safe transfer, handling and use of any living modified organism resulting from biotechnology that may have adverse effect on the conservation and sustainable use of biological diversity".

---

other self-propagating material) and envisages two different procedures enabling the Parties to make risk assessment decisions. The first procedure involves an Advance Informed Agreement (AIA) applicable to LMOs that are intended to be cultivated in the country to which they are transported. The second procedure, referring to products that are intended for use as food, feed or for processing, involves the obligation to inform a central clearing house known as the Biosafety Clearing House (BCH) (Box 6.3), of any internal decision regarding marketing permits and to provide specific information, including a relevant risk assessment, to the BCH. The AIA procedure essentially requires LMOs that are transferred between two countries for the first time to obtain a "visa" before the transfer. Risk analysis

---

BOX 6.3

THE BIOSAFETY CLEARING HOUSE

The BCH (http://bch.cbd.int) is a mechanism set up by the Protocol to facilitate exchange of information on LMOs and to assist the Parties to better comply with their obligations under the Protocol for an Information Centre on national regulations, risk assessments and final decisions and proposes a roster of experts.

It contains a wide range of biosafety information including:

- Lists of national contacts, national laws and regulations;
- Countries' decisions on GM applications;
- Registries of LMOs (events), genes and organisms;
- Roster of experts;
- Advice on capacity building (e.g., training courses);
- A scientific bibliography database.

---

then may be needed in the importing country. The structure of information flow for transboundary movement of LMOs is shown in Box 6.4.

Various articles in the Protocol outline the approaches to be used by the exporting country in making the biosafety assessment and by the importing country in using that assessment (Box 6.4).

The Protocol does not prescribe any particular type of regulatory system for the exporting country. Instead, it provides countries that do not yet have domestic biosafety legislation with a legal basis to make informed decisions regarding the safety of imported LMOs and products thereof.

3. The World Trade Organization (WTO) Agreements

a. The WTO   The WTO was established in 1995 as a successor organization to the General Agreements on Tariffs and Trade (GATT), primarily to administer the trade agreements associated with the latter. It provides a forum for trade negotiations and avails itself as a dispute settlement mechanism. A number of agreements administered by the WTO (Table 6.4) are relevant to the trade of GM-derived commodities and/or processed products, though none of these agreements mentions biotechnology specifically.

b. The Sanitary and Phytosanitary Measures (SPS) Agreement   The most relevant Agreement is that on the Application of Sanitary and Phytosanitary Measures (SPS). This recognizes that imported and domestic

BOX 6.4

THE CARTAGENA PROTOCOL

The Protocol sets up a structure for the transference of biosafety information from a country wishing to export GMO material (LMOs) to an importing country (see Fig. 1).

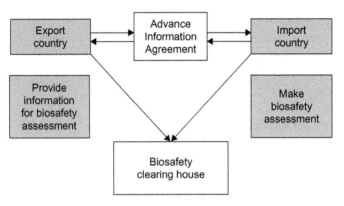

FIGURE 1   Structure for the transference of biosafety information (see colour section).

The approach that the exporting country should take to obtain the biosafety assessment information and for the importing country to assess that information is set out in various Articles in the Protocol:

Article 1 supports the Precautionary Approach (see Box 2.3) contained in Principle 15 of the Rio Declaration on Environment and Development and sets as its Objective "to contribute to ensuring an adequate level of protection in the field of the safe transfer, handling and use of living modified organisms resulting from modern biotechnology that may have adverse effects on the conservation and sustainable use of biological diversity, taking also into account risks to human health, and specifically focusing on transboundary movements".

Article 15 makes clear that the risk assessment procedure should be science based in stating "Risk assessments undertaken pursuant to this Protocol shall be carried out in a scientifically sound manner."

Annex III of the Protocol identifies the principles for scientific risk assessment that need to be addressed by member countries when considering LMOs that might have adverse effects on biological diversity, also taking into account the impact on human health. "Risk assessment should be carried out in a scientifically sound and transparent manner and can take into account expert advice of, and guidelines developed by, relevant

(cont'd)

BOX 6.4   (*cont'd*)

international organizations." It states, furthermore, that "lack of scientific knowledge or scientific consensus should not necessarily be interpreted as indicating a particular level of risk, an absence of risk, or an acceptable risk" which further enhances the Precautionary Approach.

Article 23 of the Cartagena Protocol requires public involvement in the decision-making process.

Article 26 allows for specific socio-economic issues to be taken into account in the process: "The Parties, in reaching a decision on import under this Protocol or under its domestic measures implementing the Protocol, may take into account, consistent with their international obligations, socio-economic considerations arising from the impact of living modified organisms on the conservation and sustainable use of biological diversity, especially with regard to the value of biological diversity to indigenous and local communities."

TABLE 6.4   WTO Agreements (material excerpted from a training CD-ROM on the WTO SPS Agreement prepared by the WTO Secretariat).

Agreements	Measures	Relevance to GMOs
SPS	1. Measures to protect human or animal life from additives, contaminants, toxins or disease-causing organisms in food and feed 2. Measures to protect animal or plant life from pests, diseases, or disease-causing organisms 3. Measures to protect a country from damage caused by the entry, establishment or spread of pests 4. Measures to protect human life from plant- or animal-carried diseases (zoonoses)	GM crops known to pose food and feed safety risks GM crops known to contain pathogenicity elements GM crops known to pose risks of invasiveness Not applicable for GM crops
TBT	Measures may be taken if proven to protect human, animal and environmental health.	Measures could be taken if a GM plant and/or products thereof are proven to be substantially non-equivalent with a non-modified counterpart.
	Measures must not be discriminatory and more trade restrictive than necessary.	Labelling of GMOs may fall under the TBT Agreement. If, however, a GM product is considered "like" a product in relation to a conventional product, then there no grounds for applying mandatory labelling.

(Continued)

TABLE 6.4 *(Continued)*

Agreements	Measures	Relevance to GMOs
TRIPs	Minimum level of protection for certain intellectual property (IP) rights. Inventions but not discoveries have to be patentable. Plants, animals and essential biological processes for the production of plants and animals may be excluded from patentability. IP protection is required for microorganisms, non-biological and microbiological processes.	The TRIPs Agreement may be invoked in IP protection disputes involving GM plants but not in conflicts regarding market access.
GATT	Exceptions from GATT rules can be made to protect health or the environment.	Measures could be taken if a GM crop and/or products thereof are proven to be substantially non-equivalent with a non-modified counterpart and if it can be shown that it is necessary to violate the GATT provisions in order to achieve health and environmental safety.

agricultural products need to be safe and must be devoid of risks to human, animal and plant health. For this purpose, members have the right to impose regulations protecting human and animal health (sanitary measures) and plant health (phytosanitary measures), provided that these are not applied in ways that are arbitrary and could constitute unjustifiable discrimination between countries or disguised restrictions on international trade.

Under the SPS Agreement, countries are allowed to set their own food safety and animal and plant health regulations provided that such regulations are science based and are applied only to the extent necessary for human, animal and plant health protection.

An SPS Agreement encourages Members to use international standards and guidelines. For those cases for which international standards do not exist, Members may adopt SPS measures *de novo* provided that they are scientifically justified.

The Precautionary Approach is reflected in Article 5.7 of the SPS Agreement and Members are allowed to exercise precaution provided that the measures taken are:

- Provisional (although no time limit is set);
- Adopted on the basis of "available pertinent information";
- An attempt "to obtain the additional information necessary for a more objective assessment of risk"; and
- Reviewed within a reasonable period of time.

The Agreement makes it clear that "there is a scientific justification if, on the basis of an examination and evaluation of available scientific information in conformity with the relevant provisions of this Agreement, a Member determines that the relevant international standards, guidelines or recommendations are not sufficient to achieve its appropriate level of sanitary or phytosanitary protection".

c. The Technical Barriers to Trade Agreement   The WTO Agreement on Technical Barriers to Trade (TBT Agreement) also covers health protection measures, although it is different to the SPS in scope. Countries may introduce TBT measures to prevent deceptive practices and to protect human, animal, plant and environmental health. These measures concern the description of product characteristics (e.g. composition, nutritional claims, etc.) and may require product labelling and documentation related to food safety, and/or specify quality and packaging standards and regulations. They should not be more trade restrictive than necessary and should not discriminate between "like" products (i.e., imports and domestic equivalents).

d. The Role of the WTO   The WTO will arbitrate on any problems arising from the trade of GM commodities between its members. Potential problems would include:

- The banning of GMOs and derivative products from importation and sales without adequate scientific justification that human and animal health are endangered;
- Applying testing and approval procedures that may be considered arbitrary and discriminatory and as such are used as unnecessary trade barriers;
- Applying labelling and identification requirements that may be considered to constitute trade barriers.

Labelling and identification requirements may become the subject of trade disputes. An increasing number of countries require or are considering the labelling of GM foods. Some other countries, most notably the USA as the largest exporter of grain and also including Canada and Argentina, require labelling for foods only if they are not substantially equivalent to their non-GM counterparts. Furthermore, at the international level, there is no common understanding as to what needs to be labelled (see Chapter 5, Section V).

Consumer demand in a number of major importing grain countries (e.g., EU, Japan) has led to "identity preservation" systems intended to completely segregate GM from non-GM foods. Such systems require traceability for the complete supply chain, from the seed and farm production stages to the delivery of the crop to the consumer ("from farm to fork"), and controls at each stage of the production and marketing to ensure that GM and non-GM varieties are not mixed (see Chapter 5, Section V). Absolute segregation (zero tolerance) places major organizational and economic costs on grain exporters and may be considered an unjustified barrier to trade (Bullock and Desquilbet, 2002).

In the case of trade disputes involving GMOs, the interplay between the Cartagena Protocol and the relevant WTO agreements is not clear. The legal arguments are dependent on whether the GMO is introduced in the environment, used as food or feed, or derivative products are introduced into the marketplace, and on whether the disputing parties are members to the Protocol, the WTO or both. The arguments as to which agreement would play a role in dispute settlements are beyond the scope of this book and are reviewed in detail elsewhere (see Zarrilli, 2005).

B. International Standards-Setting Bodies

The role of international standards-setting bodies and international organizations acting as facilitators of regulatory harmonization in biotechnology cannot be overstated. For example, the SPS Agreement (Article 3) encourages WTO members to base their measures on international standards, guidelines and recommendations, where they exist, and recognizes three standards-setting bodies, namely the *Codex Alimentarius* Commission (http://www.codexalimentarius.net), the Commission on Phytosanitary Measures and the Office International des Epizooties. The SPS Agreement makes no legal distinction between the "standards", "guidelines" and "recommendations" of these three organizations, all of which have equal status under the SPS Agreement. The work of the *Codex Alimentarius* Commission is particularly relevant in the context of GM food safety. The *Codex* was born out of the Joint FAO/WHO Food Standards Program with the objectives of consumer protection, harmonization of food standards developed by other international bodies, and ensuring fair trade.

An *Ad Hoc* Intergovernmental Task Force on Foods Derived from Biotechnology was established under the auspices of the *Codex* in 1999. The work of the Task Force culminated in three documents: Principles for the Risk Analysis of Foods Derived from Modern Biotechnology (Principles Document), Guideline for Safety Assessment of Foods Derived from Recombinant-DNA Plants (Plant Guideline) and Guideline for Safety Assessment of Foods Derived from Recombinant-DNA Microbes (*Codex Alimentarius*, 2003).

The Principles Document is summarized by Paoletti *et al.* (2008) (Box 6.5).

Also of particular relevance is the work of the Organization of Economic Cooperation and Development (OECD). Numerous references to this have been made elsewhere in this book and the reader is referred to the website of the organization (http://www.oecd.org/department/0,3355,en_2649_34385_1_1_1_1_1,00.html).

BOX 6.5

AD HOC INTERGOVERNMENTAL TASK
FORCE ON FOODS DERIVED FROM
BIOTECHNOLOGY OF THE *CODEX
ALIMENTARIUS* COMMISSION

Guideline Documents

This document discusses risk assessment, risk management and risk communication, and describes the safety assessment as a component of the risk assessment. The essence of the safety approach is that the new food (or component thereof) should be compared with a food already accepted as safe based on its history of safe use. The assessment should follow a structured and integrated approach. It should evaluate both intended and unintended effects, that is, intended and unintended differences from the conventional counterpart; it should identify new or altered hazards; and it should identify any changes in key nutrients that are relevant to human health. In the Guideline for the conduct of the food safety assessment of foods derived from recombinant-DNA plants the principles for risk analysis of foods derived from modern biotechnology are further detailed. For example, paragraph 4 of the Plant Guideline reiterates that rather than trying to identify every hazard associated with a particular food, a safety assessment should take a comparative approach and identify new or altered hazards relative to the conventional counterpart. Paragraph 5 of the Plant Guideline notes that if a new or altered hazard, a nutritional issue or other food safety concern is identified, one would then need to determine its relevance to human health. If all significant differences are identified and found not to pose safety concerns, then the new food can be considered to be as safe as its conventional counterpart. The framework for conducting such a safety assessment is outlined in paragraph 18 of the Plant Guideline. It states that the safety assessment of a food derived from a recombinant-DNA plant follows a stepwise process of addressing relevant factors that include:

A. Description of the recombinant-DNA plant
B. Description of the host plant and its use as food
C. Description of the donor organism(s)
D. Description of the genetic modification(s)
E. Characterization of the genetic modification(s)
F. Safety assessment:
   a. expressed substances (non-nucleic acid substances)
   b. compositional analyses of key components

(cont'd)

BOX 6.5 (*cont'd*)

c. evaluation of metabolites
d. food processing
e. nutritional modification
f. other considerations

References

Bullock, D.S. and Desquilbet, M. (2002). The economics of non-GMO segregation and identity preservation. *Food Policy* **27**, 81–99.

Codex Alimentarius (2003). *Codex principles and guidelines on foods derived from biotechnology.* Codex Alimentarius Commission Joint FAO/WHO Food Standards Programme, Food and Agriculture Organization, Rome.

Commission of the European Communities (2000). Communication from the Commission on the precautionary principle. Brussels, Belgium: Commission of the European Communities, 29 pp. (http://europa.eu.int/comm/dgs/health_consumer/library/pub/pub07_en.pdf)

Kalaitzandonakes, N., Alston, J.M. and Bradford, K.J. (2007). Compliance costs for regulatory approval of new biotech crops. *Nature Biotechnol.* **25**, 509–511.

Ochave, J.-M.A. and Estacio, J.-F.L. (2001). Biosafety regulation in the Philippines: salient points and pending issues. Paper delivered at the APEC-2001 Conference on Biotechnology, September 6, 2001, Bangkok, Thailand.

OECD (1986). Recombinant DNA safety considerations. http://dbtbiosafety.nic.in/guideline/OACD/Recombinant_DNA_safety_considerations.pdf

OECD (1992). *Safety Considerations for Biotechnology.* Organization for Economic Co-operation and Development, Paris, France.

OECD (1993). *Safety Evaluation of Foods Derived by Modern Biotechnology, Concepts and Principles.* Organization for Economic Co-operation and Development, Paris.

Paoletti, C., Flammb, E., Yanc, W., Meekd, S., Renckensa, S., Fellouse, M. and Kuiper, H. (2008). GMO risk assessment around the world: some examples. *Trends Food Sci. Technol.* **19**, S70–S78.

Pray, C.E., Bengali, P. and Ramaswami, B. (2005). The cost of biosafety regulation: the Indian experience. *Q. J. Internat. Agricul.* **44**, 267–289.

Rosendal, G.K. (2008). Interpreting sustainable development and societal utility in Norwegian GMO assessments. *European Environ.* **18**, 243–256.

UNEP-GEF (2005). Phase 0 toolkit module: starting the project. www.unep.org/biosafety/Documents/Toolkit_for_the_Development_Project_Starting_the_Project.pdf

Zarrilli, S. (2005). International Trade in GMOs and GM Products: National and Multilateral Legal Frameworks. Policy Issues in International Trade and Commodities. Study Series No. 29. United Nations Conference on Trade and Development.

Further Reading

Jaffe, G. (2004). Regulating transgenic crops: a comparative analysis of different regulatory processes. *Transgenic Res.* **13**, 5–19.

Appendix A

Commonly Used Terms

The list below defines terms in the text of this book. Further definitions of biotechnology terms can be found in http://biotechterms.org/sourcebook/index_kc.phtml and www.fao.org/DOCREP/003/X3910E/X3910E04.htm

Allele:

An allele is one alternative form of a gene (of which there may be two or more) that is located at a specific position on a specific chromosome (a locus).

Allelochemicals:

Chemicals produced by individuals of one species that modify the behaviour of individuals of a different species (i.e., an interspecific effect). These include allomones (only the emitting species benefits), synomones (both species benefit) and kairomones (the receptor species benefits).

Allopolyploids:

A polyploid in which the several chromosome sets are derived from more than one species.

Apomyxis:

The asexual production of diploid offspring without the fusion of gametes. The embryo develops by mitotic division of the maternal or paternal gamete or, in the case of plants, by mitotic division of a diploid cell of the ovule.

Backcross:

Crossing an organism with one of its parents or with the genetically equivalent organism. The offspring of such a cross are referred to as the backcross generation or backcross progeny.

Biocoenosis:

A group of interacting organisms (e.g., plants, animals, microbes) that live in a particular habitat and form an ecological community.

Biota:

All the plant and animal life of a particular region.

Chromosome:

A tightly coiled strand of DNA condensed into a compact structure by complexing with accessory histones or histone-like proteins. Chromosomes contain most of a cell's DNA. They exist in pairs in eukaryotes – one paternal (from the male parent) and one maternal (from the female parent). Each eukaryotic species has a characteristic number of chromosomes. Bacterial cells and viruses contain only a single chromosome, consisting of a single or double strand of DNA or, in some viruses, RNA, without histones.

Cisgenic:

A transgenic plant transformed with a natural gene, coding for an (agricultural) trait, from the crop plant itself or from a sexually compatible donor plant that could be used in traditional plant breeding.

Cleistogamy:

The production of flowers that do not open but instead are self-fertilized in the bud.

Construct:

A "package" of genetic material (containing more than one gene) which is inserted into the genome of a cell via GM techniques. May include promoter(s), leader sequence, termination codon, etc. Also termed a "cassette" (see Fig. 1.6).

Co-transformation:

Co-transformation is a technique in which recipient cells are incubated with two types of plasmid, one of which is selectable and the other not. Cells which have been transformed with the first plasmid are then selected. If transformation has been carried out at a high DNA concentration, then it is probable that these cells will also have been transformed with the second (non-selectable) plasmid.

DNA (deoxyribonucleic acid):

The long chain of molecules in most cells that carries the genetic information (genes) and other elements; it controls all cellular functions in most forms of life. DNA is a macro-molecule composed of a long chain of deoxyribonucleotides joined by phosphodiester linkages. Each deoxyribonucleotide contains a phosphate group, the five-carbon sugar 2-deoxyribose, and a nitrogen-containing base. The genetic material of most organisms and organelles so far examined is double-stranded DNA, though a number of viral genomes consist of single-stranded DNA. In double-stranded DNA, the two strands run in opposite (anti-parallel) directions and are coiled round one another in a double helix. Purine bases (adenine (A) and guanine (G)) on one strand specifically hydrogen bond with pyrimidine (cytosine (C) and thymine (T)) bases on the other strand, according to the Watson–Crick rules (A pairs with T; G pairs with C).

Edaphic:

In ecology, edaphic refers to plant communities that are distinguished by soil conditions rather than by the climate.

Event:

Refers to each instance of a genetically engineered organism, that is each instance of novel genetic material being introduced into a recipient organism through GM techniques. Generally speaking, regulatory agencies confer new biotech-derived product approvals on an event-by-event basis.

Excision marker genes:

Systems that eliminate selectable marker genes by strategies such as site-specific recombination, homologous recombination or transposition from the genome after they have fulfilled their purpose. An example is the Cre-Lox site-specific DNA recombinase system.

Expression:

The translation of the cell's genetic information stored in the DNA (gene) via RNA (transcription) into a specific protein (synthesized by the cell's ribosome system). In the case of transgene expression, it is the production of a novel protein encoded by the introduced genetic material.

Gene:

The unit of heredity transmitted from generation to generation during sexual or asexual reproduction. Typically, a gene is a segment of nucleic acid that is transcribed to RNA and is then expressed as a peptide or protein. It comprises a promoter which informs the cellular machinery where to start transcription of the RNA, a coding sequence for the protein, and a termination sequence which stops the transcription. Often there are additional DNA sequences which control factors such as the tissue specificity, timing and level of expression.

The term is also used for sequences that express RNA for RNA (gene) silencing and which do not produce a protein.

Gene silencing:

The suppression of gene expression (e.g., of the gene for polygalacturonase which causes fruit to ripen, of the gene for P34 protein in soybeans, etc.) via a variety of methods (e.g., via RNA interference (RNAi), chemical genetics, effect of certain viruses, "zinc finger proteins", sense or antisense genes, etc.). Also occurs with some genes in an organism as the organism matures (e.g., from an embryo to a seedling/juvenile). Can also be an unintended effect resulting from introducing genetic material with duplicated elements.

Genotype:

The genetic constitution (gene makeup) of an organism.

Heterozygote (*adjective* heterozygous):

An individual that has different alleles at the same locus on its two homologous chromosomes.

Homozygote (*adjective* homozygous):

An individual that has two copies of the same allele at a particular locus on its two homologous chromosomes.

In-breeding/out-breeding:

In-breeding is mating between individuals that have one or more ancestors in common, the extreme condition being self-fertilization which occurs in many plants. In-breeding contrasts with out-breeding, which is a mating system characterized by the breeding of genetically unrelated or dissimilar individuals.

Isogenic:

Adjective describing organisms that are genetically identical; completely homozygous.

Isoline:

An in-bred line that is completely homozygous. Also used to describe a line that is genetically identical to a genetically transformed line apart from the introduced genetic elements.

Landrace:

An early, cultivated form of a crop species derived from a wild population and typically containing unique locally adapted phenotypes.

Line:

A group of identical or near-identical pure-breeding diploid or polyploid organisms, distinguished from other individuals of the same species by some unique phenotype and genotype.

Marker-assisted breeding (also termed marker-assisted selection):

The use of DNA sequences (markers) specific to individual traits to select individuals in a breeding population. Its use can speed up conventional breeding as the selection can be made before the trait becomes apparent.

Meiosis:

The cell division process by which the chromosome number of a reproductive cell is reduced to half (n) the diploid (2n) or somatic number. Two consecutive divisions occur. In the first division, homologous chromosomes become paired and may exchange genetic material (via crossing over) before moving away from each other into separate daughter nuclei (reduction division). These new nuclei divide by mitosis to produce four haploid nuclei. Meiosis results in the formation of sexual cells (gametes) in plants and animals. It is an important source of variation through recombination.

Mutation:

An instantaneous heritable change appearing in an individual as the result of a change in the structure of a gene (=gene mutation), changes in the structure of chromosomes (=chromosome mutation), or in the number of chromosomes (=genome mutation).

Phenotype (*adjective* phenotypic):

The visible appearance or set of traits of an organism resulting from the combined action of genotype and environment. Cf. *genotype*.

Plasmid:

An extrachromosomal circular DNA molecule found in certain bacteria, capable of autonomous replication. Plasmids can transfer genes between bacteria and are important tools of transformation in genetic engineering. They exist in an autonomous state and are transferred independently of chromosomes.

Pleiotropy (*adjective* pleiotropic):

The situation in which a particular gene has an effect on several different traits.

Polyploid:

Cells and organisms that have more than two homologous sets of chromosomes.

Pre-/post-zygotic:

Before (pre-) or after (post-) a zygote is formed; occurring before or after completion of fertilization.

Promoter:

A nucleotide sequence of DNA to which RNA polymerase binds, leading to the initiation of transcription. It usually lies upstream of (5' to) a coding sequence. A promoter sequence aligns the RNA polymerase so that transcription will initiate at a specific site.

Site-specific recombination:

A type of genetic recombination in which DNA strand exchange takes place between segments possessing only a limited degree of sequence homology (typically between 30 and 200 nucleotides in length). Site-specific recombinases rearrange DNA segments by recognizing and binding to short DNA sequences, at which they cleave the DNA backbone, exchange the two DNA helices involved and rejoin the DNA strands. Site-specific recombination systems are highly specific, fast and efficient, even when faced with complex eukaryotic genomes. They are employed in a variety of cellular processes, including bacterial genome replication and movement of mobile genetic elements. For the same reasons, they present a potential basis for the development of genetic engineering tools.

Species:

A class of potentially interbreeding individuals (genetically compatible) that are reproductively isolated from other such groups and have many characteristics in common. It is a somewhat arbitrary classification, but still useful in most situations.

Termination sequence:

A DNA sequence just downstream of the coding segment of a gene, which is recognized by RNA polymerase as a signal to stop synthesizing mRNA. There may also be sequences beyond the transcribed part of the gene which influence the termination of transcription.

Totipotent:

Having the potential to form all the types of cells in an organism.

Trait:

A characteristic of an organism, which manifests itself in the phenotype (physically). Many traits are the result of the expression of a single gene, but some are polygenic (result from expression of more than one gene). For example, the level of protein content in soybeans is controlled by five genes. In genetic modification, it refers to the phenotype produced by the newly introduced gene(s).

Transformation (*noun* transformant):

The uptake and establishment of exotic DNA in a bacterial, yeast, plant or animal cell, in which the introduced DNA often changes the phenotype of the recipient organism.

Transposon:

A transposable or movable genetic element. A relatively small DNA segment that has the ability to move (mobile genetic element) from one chromosomal position to another; for example Tn 5 is a bacterial transposon that carries the genes for resistance to the antibiotics neomycin and kanamycin and the genetic information for insertion and excision of the transposon.

V-GURTs:

V-GURTs are a type of Genetic Use Restriction Technology (GURT), colloquially known as "terminator" technology. GURT is the name given to proposed methods for restricting the use of GM plants by causing second generation seeds to be sterile. There are conceptually two types of GURT: (1) V-GURTs which produce sterile seeds of that variety (V), meaning that a farmer that had purchased seeds containing V-GURT technology could not save the seed from this crop for future planting; and (2) T-GURTs in which a crop is modified such that the genetic trait (T) engineered into the crop does not function until the crop plant is treated with a chemical. In the latter case, farmers can save seeds for use each year, but do not get to use the enhanced trait in the crop unless they purchase the activator compound.

Volunteer:

A plant from a previous season's crop that regenerates in a subsequent crop, for example potato tuber or shed seed. Typically undesirable because of effects on the management and yield of the subsequent crop.

Appendix B

Information Required on an Application for Environmental Release of a GMO

Below is outlined the type of information that is required in an application for release of a GMO in a specific country (X). The form is for environmental release and a different one would be required for assessing food and feed safety. It is based on the Canadian application forms* and is intended as guidance to constructing an application form. It is fully recognized that the regulatory authority of country X can design its own application system and that there may be need for other information depending on factors such as country and region. See case studies in Appendix D for further examples.

As noted above, this guidance is for applications in Canada. Not all the questions to be addressed may be relevant to Country X, and it is for the regulatory authorities in that country to decide how much information they require.

1 Personnel involved and status of the GMO in the application
1.1 Applicant:

1. Name
2. Address
3. Telephone number
4. Facsimile number
5. E-mail address

1.2 National representative, if different from above:

1. Name
2. Address
3. Telephone number

---

* Sources http://www.inspection.gc.ca/english/plaveg/bio/dir/dir9408e.shtml#ch8-2; http://www.inspection.gc.ca/english/plaveg/bio/dir/dir9408appe.shtml

---

4. Facsimile number

5. E-mail address

1.3 Is the plant material imported? If yes, was an import permit applied for under the *relevant national system*? Was it granted? If yes, provide the permit number if known.

1.4 Was the plant material previously tested/grown/released in country X? If yes, in what years?

1.5 If the GMO was derived through recombinant DNA techniques, were the gene constructs previously tested in country X? If yes, in what plant species and in what years?

1.6 Were other government agencies, either foreign or within country X, notified of the development of the GMO or its importation? What was the purpose of such notification?

2 Biology of the recipient plant species

The biology of certain plant species is described in a series of species-specific biology documents, an example of which can be found at: http://www.inspection.gc.ca/english/plaveg/bio/dir/biodoce.shtml.

These documents describe the characteristics of the plant species in question, such as habitat, fertility, dispersal and endogenous toxins, as well as include information about the plant species' major interactions with other life forms in its production range in Canada (e.g., predators, grazers, parasites, pathogens, competitors, symbionts and beneficial organisms, including humans, where appropriate). This information will help identify potential risks associated with a GMO under review relative to its counterpart(s) of the same species already present in country X's environment. These documents act as references for comparative data.

Where a biology document for a particular GMO's plant species is not available, applicants are strongly recommended to notify the regulatory authority at least six months prior to the anticipated submission of an application for unconfined environmental release. This will allow adequate time for the development of the new biology document. This document will be drafted using subject matter experts, published peer-reviewed literature, and consensus documents developed by the Organization for Economic Cooperation and Development (OECD). Please note that the review of an application for a GMO will not be initiated until a finalized biology document is available. Therefore, it is in an applicant's best interest to notify the regulatory authority as early as possible if a required biology document is not available in order to avoid any delays in the assessment of their application. The format of the new biology document will follow that of the existing biology documents.

3 Description of the GMO

3.1 Taxonomy and use

3.1.1 Description of its taxonomy.

3.1.2 Designation given to the GMO, including all synonyms. Each GMO line should have a unique identifier designated according to the "OECD Guidance for the Designation of a Unique Identifier for Transgenic Plants" (ENV/JM/MONO(2002)7) (http://www.olis.oecd.org/olis/2002doc.nsf/LinkTo/env-jm-mono(2002)7).

3.1.3 Pedigree information of the GMO (including any relationship to a previously assessed GMO).

3.1.4 Give details of the use of the GMO (e.g., to be grown as a field crop for grain production; to be grown as a field crop for grain production on lands contaminated with persistent herbicide; to reclaim lands contaminated with heavy metals).

3.2 Description of the modification of the GMO

3.2.1 Identify the objective of the modification, e.g., novel herbicide tolerance, male sterility/restoration, etc.

3.2.2 Describe and provide references for the transformation method, e.g., Agrobacterium-mediated transformation or direct transformation by methods such as particle bombardment, electroporation, polyethylene glycol transformation of protoplasts, etc.

3.2.3 For direct transformation methods, describe the nature and source of any carrier DNA used.

3.2.4 For Agrobacterium-mediated transformation, provide the strain designation of the Agrobacterium used during the transformation process, and indicate if tumor-inducing genes were present on the plasmid-based vector, and whether Agrobacterium was cleared from the transformed tissue. Briefly describe or provide reference(s) describing the construction of the vector.

3.2.5 For transformation systems other than Agrobacterium, provide the following information:

1. Does the system utilize a pathogenic organism or nucleic acid sequences from a pathogen?
2. How were any pathogenesis-related sequences removed prior to transformation?
3. Did the transformation process involve the use of helper plasmids or a mixture of plasmids? If so, describe these in detail.

3.2.6 Description of the genetic material potentially delivered to the recipient plant material (the modification/constructs).

3.2.7 Provide a summary of all genetic components which comprise the vector including coding regions, and non-coding sequences of known function. An example of a table describing the DNA components of a vector (from APHIS petition #94-257-01P) is available at http://www.inspection.gc.ca/english/plaveg/bio/usda/usda03e.shtml#table1. For each genetic component provide a citation where these functional sequences were described, isolated and characterized (publicly available database citations are acceptable) and indicate:

a. The portion and size of the sequence inserted.
b. The location, order and orientation in the vector.
c. The function in the plant.
d. The source (scientific and common, or trade name, of the donor organism).
e. If the genetic component is responsible for disease or injury to plants or other organisms, and is a known toxicant, allergen, pathogenicity factor, or irritant.
f. If the donor organism is responsible for any disease or injury to plants or other organisms, produces toxicants, allergens or irritants or is related to organisms that do.
g. If there is a history of safe use of the source organism or components thereof.

3.2.8 If there has been a modification in the transgene relative to the native gene that affects the amino acid sequence of the protein designed to be expressed in the plant, provide the citation. If the modified amino acid sequence has not been published, provide the complete deduced sequence highlighting the modifications. Indicate whether the modifications are known or expected to result in changes in post-translational modifications or sites critical to the structure or function of the gene product. An example of such modifications might include the addition of new glycosylation sites.

3.2.9 Provide a detailed map of the vector with the location of sequences described above that is sufficient to be used in the analysis of data supporting the characterization of the DNA, including as appropriate the location of restriction sites and/or primers used for PCR and regions used as probes. An example of a detailed map of a plasmid vector (from APHIS petition #94-257-01P) is available at the following address: http://www.inspection. gc.ca/english/plaveg/bio/usda/usda03e.shtml#figure

3.3 Characterization of the DNA inserted in the plant

3.3.1 For all coding regions, provide data that demonstrate if complete or partial copies are inserted into the plant's genome. Coding regions may include truncated sense constructs, sequences engineered to be non-translatable, antisense constructs, and constructs containing ribozymes, regardless of whether or not the coding region is designed or expected to be expressed in the transgenic plant. For allopolyploid plants, information may be required indicating into which parental genome insertion has occurred.

3.3.2 For non-coding regions associated with the expression of coding regions:

a. Data should demonstrate whether or not plant promoters are inserted intact with the coding regions whose expression they are designed to regulate. These data are relevant to consideration of points 3.4.1 and 3.4.2 below.

b. DNA analysis may be necessary for introns, leader sequences, terminators and enhancers of plant-expressible cassettes. DNA analyses may be presented in the form of Southern analyses, DNA sequencing, PCR analyses, or other appropriate information.
c. DNA analysis may be necessary for promoters and other regulatory regions associated with bacteria-expressible cassettes.

3.3.3 For non-coding regions which have no known plant function and are not associated with expression of coding regions:

a. DNA analysis may be required for some sequences of known function (e.g., ori V and ori-322, bom, T-DNA borders of Agrobacterium, and bacterial transposable elements).
b. DNA analysis is not necessary for any remaining sequences of the plasmid backbone when the plasmid is well characterized.

3.3.4 Where appropriate, provide sequence data of the inserted material and of the surrounding regions (sequencing information may be informative in some cases, that is to fully characterize a partial or rearranged DNA insert).

3.4 Protein and RNA characterization and expression

3.4.1 For all complete coding regions inserted, provide data that demonstrate whether the protein is or is not produced as expected in the appropriate tissues consistent with the associated regulatory sequences driving its expression (e.g., if the gene is inducible, determine if the gene is expressed in the appropriate tissues under induction conditions). For virus-resistant plants where the transgenes are derived from a viral genome, in addition to transgene protein analysis, determine transgene RNA levels in tissues consistent with the associated regulatory regions driving expression of the transgene. The following exceptions also apply:

a. If the protein concentration is below the limits of detection, mRNA data may be substituted.
b. Protein analysis for products of genes used only as selectable markers may be waived under certain circumstances, e.g., when there is at least one complete copy of a selectable marker gene present and the effective expression of the selectable marker gene is verified by the process used to select the transformed tissue.
c. For plants modified to express non-translatable mRNA, truncated sense constructs, antisense constructs, or constructs containing ribozymes, since the function of these genetic constructs is to specifically alter the accumulation of a specific mRNA or protein present in the transgenic plant, provide data on the level of the target protein only (e.g., native tomato fruit polygalacturonase would be the target protein of antisense polygalacturonase to achieve altered fruit ripening). If the target protein levels are below levels of detection, determine target mRNA levels.

3.4.2 When a fragment of a coding region designed to be expressed in a plant is detected, determine whether a fusion protein could be produced and in which tissues it may be located.

3.4.3 Protein or RNA characterization may not be required for fragments of genetic constructs not expected to be functional in the plant (e.g., fragments of selectable marker genes driven by bacterial promoters).

3.5 Description of the inheritance and stability of introduced traits which are functional in the plant

3.5.1 For plants which are either male or female fertile, or both, provide data that demonstrates the pattern and stability of inheritance and expression of the novel traits. If the new trait cannot be directly measured by an assay, it may be necessary to examine the inheritance of the novel DNA sequences directly, and expression of the RNA.

3.5.2 Plants which are either infertile or for which it is difficult to produce seed (such as vegetatively propagated male-sterile potatoes) provide data to demonstrate that the novel trait is stably maintained and expressed during vegetative propagation over a number of cycles that is appropriate to the plant.

3.6 Description of the parental genome

In the case of an allopolyploid in which parental genome is the genetic modification?

3.7 Number of generations removed from the original modification

3.8 Description of the novel traits

3.8.1 Where applicable, characterize in detail the novel gene products, breakdown products, by-products and their metabolic pathways.

3.8.2 Is the novel trait expressed in a tissue-specific manner?

3.8.3 Is the novel trait expressed in a developmental stage-specific manner?

3.8.4 Is expression of the novel trait induced? If yes, what are the inducing agents?

3.8.5 Where applicable, describe the activity of the gene products, breakdown products and by-products in the host plant. Describe any changes to existing metabolic pathways (including altered accumulation and storage patterns), including those that might not be intended.

3.8.6 Where applicable, the toxicity of the novel gene products, breakdown products and by-products in the environment must be established. Describe:

a. potential toxigenicity to known or potential predators, grazers, parasites, pathogens, competitors and symbionts;
b. potential for adverse human health effects, e.g., exposure to toxins, irritants and antigens. Include estimated level and most likely route of human exposure to the gene products, breakdown products and by-products.

4 Environmental characterization

Note that data from at least two seasons of trials in multiple locations in country X or in a similar environment are normally required to address these questions.

4.1 Selection of counterpart

Generally, the most suitable counterpart for comparative studies is the isogenic line closest to the GMO, provided that the GMO is intended to be cultivated in the same region as this line. The counterpart may be a previously authorized GMO that has been in large-scale commercial production for several years. Developers are encouraged to consult with the regulatory authority where there are questions regarding selection of an appropriate counterpart.

4.2 Phenotype of the GMO

The applicant must provide information on the intended phenotype and any known unintended or unanticipated traits. The GMO should be compared to its counterpart(s) and related cultivated varieties as appropriate. If differences are detected, the applicant should address these findings in the application.

Typically, observations are made when the plants are grown in multiple sites and over more than one growing season. Confined research field trials of GMOs should take place in the intended growing region of the GMO in country X. Data collected from field studies outside country X can be used if the applicant demonstrates that the environment for testing the GMO is similar to country X's environment. In some cases, such as where there may be a potential for increased weed characteristics or if the plant is an out-crossing species, it may be appropriate to evaluate the plants outside of managed ecosystems. Depending upon the results, additional studies may be warranted to provide the required information. Applicants may provide valid scientific rationale why certain information is unnecessary or inappropriate.

4.2.1 Describe the breeding history of the GMO population being evaluated starting at the point of trait introduction.

4.2.2 Compare the GMO to its counterpart with respect to the following characteristics which influence reproductive and survival biology:

a. Habit – Note any changes in basic morphology of the plant including any abnormalities, e.g., changes in overall growth habit, pollen characteristics (such as stickiness, size), seed shattering dormancy characteristics, symbionic associations (e.g., with vesicular–arbuscular mycorrhizal fungi, rhizobia), etc.

b. Life cycle – e.g., plants are categorized as annual, biennial, perennial. Would the presence of the introduced trait produce a change?

c. Life history characteristics such as:
   i. plant height, biomass (dry/wet weight)
   ii. number of flowers produced/plant

   iii. number of fruits produced/plant
   iv. number of pollen grains/anther
    v. percentage of viable pollen
   vi. time to maturity (e.g., time to flowering)
  vii. number of viable seeds produced/fruit
 viii. percentage of seed germination
   ix. percentage of germinated seeds surviving to maturation
    x. number of flowering days

d. Out-crossing frequency (within species) and/or cross-fertilization frequency (between species).
e. Impact on pollinator species – this may be addressed through information on whether the same pollinator species have been seen in the field or have there been changes in the pollinators that visit the flowers (requires previous familiarity with pollinators on non-GMO species).
f. Stress adaptations (specifically note which stresses were observed):
  i. Biotic stress factors: Examples might include parasites or pathogens, competitors (e.g., weeds), and herbivores.
  ii. Abiotic stress factors: Examples might include response to drought stress, nutrient deficiency or other stresses common to that plant.
g. Ability to overwinter (or overseason).

4.2.3 Compare the compositional analysis of the GMO to its counterpart(s) including protein, lipids, fibre, and other parameters as appropriate. These data are used to assess secondary or pleiotropic effects and may indicate environmental impacts (e.g., changes in nutritional quality of seeds affecting birds).

4.2.4 Compare the GMO and its counterpart(s) with respect to levels of known naturally expressed toxicants, antinutrients and allergens known for that species.

4.3 Cultivation of the GMO

4.3.1 Description of area of cultivation:

a. Describe the regions in country X where the species is currently cultivated.
b. Will the modification permit cultivation of the species in regions in country X outside the area of current cultivation? If so, in what new regions might the GMO be cultivated? Describe the ecosystems in the new regions.

4.3.2 Description of cultivation practices:

a. Describe the cultivation practices for the GMO, including land preparation, fertilizer usage, weed and pest control, harvest, post-harvest protocols, and other cultivation practices. Compare and contrast these practices with those traditionally used for this species. Discuss how such practices might influence agro-ecosystem sustainability, crop rotations, pesticide use, frequency of tillage, soil erosion and consequential changes in energy and soil conservation. Discuss in what ways any volunteer plants of the GMO may dictate altered management practices for succeeding crops?

b. Describe any specific deployment strategies recommended for this GMO? Deployment strategies might include geographic or temporal factors or integration with other practices.

   i. Insect Resistance Management – In the case of insect resistant GMOs, describe strategies intended to delay the development of resistance in target insect populations.

   ii. Herbicide Tolerance Management – In the case of GMOs developed for tolerance to a herbicide or class of herbicides, describe appropriate strategies that are intended to delay the development of herbicide tolerance weeds and avoid significant changes in weed biotypes.

4.4 Interactions of the GMO with sexually compatible species

Determine whether there are any sexually compatible species in areas where the GMOs will be grown. If there are, then this section is applicable and the following questions should be considered.

4.4.1 Which sexually compatible species, if any, are found in areas where the plant will be cultivated, including any new areas of cultivation?

4.4.2 Characterize the compatible wild relative(s) with respect to weediness in managed ecosystems and/or establishment and spread into natural ecosystems.

4.4.3 In what ways would the introduced trait itself be likely to change the ability of the GMO to interbreed with other plant species?

4.4.4 In cases in which there is a potential for gene flow from the GMO into sexually compatible species (e.g., same or related species as appropriate), describe the consequences for the offspring of such crosses. Characterization of the crosses between wild relatives and GMOs should be considered using the criteria described above for GMOs in order to address questions (a) and (b) below.

a. Is the introduced trait similar to a trait found currently in natural populations of the compatible wild relatives?

b. Does the introduced trait have the potential to increase the reproductive fitness or confer a selective advantage on the wild relative? If so, would the introduced trait have a significant impact on the establishment and spread of populations of wild relatives? Consider the presence or absence of selection pressures.

   i. Is the potential for the trait to increase reproductive fitness or confer a selective advantage different than the potential for this to occur from a similar trait that may already exist for the same plant?

4.5 Residual effects and toxicity on non-target organisms

4.5.1 Characterize the extent to which the gene product has been a part of the human or animal diet.

4.5.2 Where applicable, characterize to what extent the introduced DNA directly or indirectly leads to the expression or altered expression of a toxin or other product that is known to affect metabolism, growth, development, or reproduction of animals, plants, or microbes.

4.5.3 Consider potential physiological and behavioural effects to other organisms including insect, avian, aquatic, or mammalian species in the areas where the plant will be cultivated, including any new area of cultivation.

Consideration may be given to:

- Threatened and endangered species in the area where the plant is to be grown.
- Beneficial organisms (pollinators, predators, parasites, biological control organisms, soil microbes).
- Other appropriate non-target organisms.

Consider levels and routes of exposure to all plant parts that express the gene, i.e., direct feeding or other exposure to the plant or plant part, dispersed plant parts, secretion, degradation, or leaching of the novel gene product, gene introgression, or organisms that have fed on the plant.

To address the possibility of persistent toxins in the environment or persistent changes in soil ecosystem function, applicants are encouraged to undertake residual effects studies. The residual effects of the GMO in comparison to the counterpart may be assessed by crop rotation studies or other techniques. Direct measurements in soil microbial communities may be indicated if microbial toxins are expected in root exudates.

4.5.4 Characterize potential adverse effects on the health of humans (including workers, adults, and children) which may arise through physical contact with or use of the GMO or its parts or its raw or processed products, other than for uses for which other authorizations or reviews are required (e.g., food, feed, pharmaceuticals). The analysis might include a comparison of the transgenic and non-transgenic counterpart(s) with respect to the likely exposure to toxins, irritants, and allergens.

4.6 Other environmental interactions

In the case of GMOs developed using plant viral coding regions, address synergy, facilitated movement, transcapsidation, and viral recombination. For a description of these terms, see the OECD "Consensus Document on General Information concerning the Biosafety of Crop Plants Made Virus Resistant through Coat Protein Gene-Mediated Protection" No. 5, 1996, OCDE/GD(96)162 (http://www.olis.oecd.org/olis/1996doc.nsf/LinkTo/ocde-gd(96)162).

5 Detection and identification

Have you provided a detection method capable of distinguishing your GMO from other commercial cultivars of the same species?

5.1 Detection and identification requirements

Along with all other data, relevant to the environmental safety assessment of a GMO, the following should also be submitted:

- Appropriate test methodologies for the detection and identification of GMOs;
- Written agreement to provide the regulatory authority with reference material suitable to support these methods.

6 Food and feed use approvals

If your GMO is intended for food and/or feed use, have you applied to the relevant national authority/authorities for food and/or feed use approval as appropriate?

GMOs that could reasonably be expected to be used as feed and food will not be authorized for unconfined environmental release by the national authority, among other requirements, until:

- The Feed Section of the national authority is ready to authorize the novel feed for livestock feed use and/or
- The national authority is ready to provide notification of no objection for human food.

Where products are intended for exclusive use as either food, feed or molecular farming (use of plants to produce industrial or therapeutic products), consultations among regulatory authorities will be required to assess any potential risks associated with the release of the product in an unintended commodity stream. For these products, an identity preservation system or alternative will be essential to minimize the likelihood of such an event.

7 Special crop management considerations

If your GMO carries a novel insect resistance, have you provided an appropriate Insect Resistance Management (IRM) plan (see Section 7.1.1)? If your GMO carries a novel herbicide tolerance, have you provided an appropriate Herbicide Tolerance Management (HTM) plan (see Section 7.1.2)? You are encouraged to work with seed distributors, extension personnel and growers to develop and implement an appropriate HTM plan for your GMO.

Please note that the regulatory authority's decision with regards to authorizing the environmental release of a GMO expressing either a novel insect resistance or a novel herbicide tolerance will take into consideration whether or not the applicant has provided a stewardship plan addressing the need for the responsible deployment of the novel crop in question into the environment.

7.1 Stewardship plan requirements

As part of the regulatory authority assessment of a GMO's environmental safety, in particular, of its assessment of longer term environmental effects, the decision with regards to authorizing the release of a GMO expressing either a novel herbicide tolerance or a novel insect resistance will take into consideration whether or not the applicant has provided a stewardship plan addressing the need for the responsible deployment of the novel crop into the environment.

Stewardship plans should include appropriate strategies that will allow for the environmentally safe and sustainable deployment of such novel plants. In addition, communication to growers and an efficient mechanism allowing growers to report problems to the applicant are all integral parts of a stewardship plan.

7.1.1 Insect resistance management (IRM)

The regulatory authority strongly recommends that an IRM plan be implemented for all plants expressing novel insect resistance (including those expressing *Bacillus thuringiensis* (Bt) endotoxins) grown in fields of **greater than one hectare in size**. IRM strategies are intended to delay the development of resistance in the insect to the active compound(s) and thereby prolong the lifespan and usefulness of the technology. The development of resistance in insects to these active novel compounds due to the non-adoption of effective IRM plans could also have significant implications on sustainable agriculture.

The IRM plan should take into consideration the most recent available scientific evidence on, among other factors, the following:

a. The reproductive biology and behaviour of the insect pest;
b. The mobility of the larvae;
c. The ability of adults to disperse from the natal field before and after mating;
d. An estimate of resistance allele frequency in the insect population;
e. The impact of management practices such as insecticide use in the refuge;
f. The targeted life cycle stage of the insect pest; and
g. Any history of insect resistance to the active compound(s).

The IRM plan submitted in an application for unconfined environmental release authorization is specific to the target insect species and is based on field/laboratory research and computer models.

Communication to growers, the monitoring of the effectiveness of the plan, and an efficient mechanism allowing growers to report problems to the applicant, are all integral parts of an IRM stewardship plan.

7.1.2 Herbicide tolerance management (HTM)

The development of an HTM plan is the applicant's responsibility and should contain elements that address:

a. The control of volunteers, more specifically, any changes in usual agronomic practices that may arise from the novel herbicide tolerance and which could result in reduced sustainability or have significant impacts on soil conservation;
b. The selection of herbicide tolerance in weeds resulting from the potential continued application of the same herbicide in subsequent rotations;
c. The introgression of novel trait into related species;
d. The management of the herbicide tolerant crop during the growing season, particularly where multiple herbicide tolerances, due to cross pollination, could arise in subsequent growing seasons;
e. Communication to growers as well as an efficient mechanism allowing growers to report problems to the developer;
f. The monitoring of effectiveness of the stewardship plan.

A GMO with a novel herbicide tolerance that could be introgressed to related species, resulting in hybrids that have no effective or sustainable control options, will not be authorized.

8 Post-release monitoring plan

You must provide a general plan for post-release monitoring of environmental effects of your GMO.

8.1 Post-release monitoring plan

A general post-release monitoring plan to monitor for unintended or unexpected environmental effects of an authorized product should also be an integral part of a complete application and will be reviewed during the environmental safety assessment of the novel plant in question. The use of appropriate indicators to evaluate these effects should be based on the characteristics of the GMO. A stewardship plan (see Section 7.1) may be considered acceptable for post-release monitoring purposes.

The applicant must inform the regulatory authority of any new information regarding the risks to the environment or to human health resulting from worker exposure to the GMO that could result from the unconfined release of the GMO.

Appendix C

Information Required for Risk Assessment

This summarizes the range of information usually required for a risk assessment and complements the draft guidance to applicants outlined in Appendix B.

Focus of risk assessment	Information required for risk assessment
The recipient plant	Information relating to the recipient (reproductive biology, gene flow and seed dispersal potential, survivability, geographic distribution)
Genetic modification method	Description of the method of genetic modification including the nature and source of vector(s); source of donor DNA and characterization of the construct
The donor organism	Information relating to whether it is known to be a pathogen, contain pathogenic sequences and/or contain sequences coding for toxins and/or allergens
GM plant – general properties	Inserted and/or deleted sequences; insert expression; phenotypic differences between the GM and the parent plants (reproduction, dissemination, survivability); genetic stability of the insert and phenotypic stability of the GM plant
Comparison of the composition of the GM plant to that of an appropriate reference with a history of safe use	Content range of macronutrients, micronutrients (vitamins, minerals), anti-nutrients and toxins of the parts of the plant parts that are consumed
Transgenic protein levels	The edible tissues of the GM plant and an appropriate non-GM control are measured for the presence of the transgenic protein

(Continued)

Focus of risk assessment	Information required for risk assessment
GM plant – food and feed safety	
Toxic properties of the transgenic protein and of the source of the transgene coding for the transgenic proteins	Data from literature, risk assessments and case reports
Comparison of the transgenic protein with proteins known to be toxic	Degree of structural similarity between the transgenic protein and known toxic proteins
Stability of the transgenic protein during gastrointestinal passage	*In vitro* digestibility of the transgenic protein incubated with proteolytic digestive enzymes under simulated conditions
Processing stability	Stability of the protein under conditions representing food processing, including elevated temperature, fermentation, etc.
In vivo toxicity of the protein	Laboratory animal studies in which the safety of the protein has been investigated
In vitro and *in vivo* toxicity of a pesticide and its metabolites formed in GM plants	Data on the occurrence and metabolism of residues of pesticides applied to GM plants, as well as their metabolism and toxicity in humans and animals
In vivo toxicity of the whole product	Laboratory animal feeding study in which the animals receive the whole GM product as part of their diet
Known allergenic properties of the source of the transgene	Data from literature and other sources regarding allergic reactions towards the source of the transgene, in particular of allergenic proteins
Comparison of the transgenic protein with known allergenic proteins	Degree of structural similarity between the transgenic protein and known allergenic proteins
Susceptibility of the protein to degradation under digestive conditions of the gastro-intestinal tract	*In vitro* digestibility of the transgenic protein (same as for toxicity, see above)
Potential allergic response in allergic patients	Cross-reactivity of the transgenic protein with allergenic proteins to which patients are allergic
Potential changes in intrinsic allergenicity of crops that have a history of allergenicity, caused by the genetic modification	Changes in the levels and/or profiles of allergenic proteins present within the host crop with a history of allergenicity
Changes in the levels and/or profiles of allergenic proteins present within the host crop with a history of allergenicity	Data on the presence of the antibiotic resistance, the clinical importance of the antibiotic and the likelihood of transfer, particularly to pathogenic micro-organisms (besides data on the safety of the transgenic protein *per se*, see above)
Changes in the levels of nutrients	Data on the levels of nutrients

(Continued)

Focus of risk assessment	Information required for risk assessment
Impact of the GM crop on animal nutrition	Data from animal trials with the GM plant, including performance or balance studies, in case of altered levels of nutrients or their altered bioavailability
GM plant – environmental safety	
Selective advantage or disadvantage	Field studies at different geographic sites in different seasons of plant growth and development, fertility, seed viability
Potential for gene transfer	Information on the reproductive biology of the plant and, in addition, of any evidence of horizontal gene transfer to other organisms
Interactions between the GM plant and target organisms	*In vivo* and *in vitro* studies of the effects of the GM plant on pest and disease organisms using a tiered approach
	Data on the comparative susceptibility of the GM plant to pests and diseases compared with non-modified plants
Interactions of the GM plant with non-target organisms	*In vivo* and *in vitro* studies of the effects of the GM plant on non-target organisms, using a tiered approach
Effects on biogeochemical processes	Studies on impacts on soil microbial communities. The fate of any (newly) expressed gene products and derivatives which result in exposure of non-target organisms in relation to the decomposition processes
	An analysis of shifts in populations of deleterious organisms
Impacts of the specific cultivation, management and harvesting techniques	Study of the commercial management regimes for the GM crop including changes in applications of plant protection products, rotations and other management measures that are different from the equivalent non-GM plant
Potential interactions with the abiotic environment	Exposure to a range of environmental conditions
Post-market environmental monitoring: case specific monitoring	Studies of the environmental factors most likely to be adversely affected by the GM plant identified in the environmental risk by studying the commercialization or exploitation of the GMO

Appendix D

Case Studies

In this appendix we give five case studies on applications to release GM crops to the field. These case studies have been selected to demonstrate different issues raised by the GM product and the general approach to making and assessing an application. In practice each application gives more details and data on the product which we summarize. We give examples of some of the data supplied and references are given for the full application.

These case studies have three aims:

- To illustrate the types of information required for a risk assessment;
- To discuss specific cases;
- To provide material for student or training exercises.

As well as the specific issues for each case study, there are common elements to them all which need to be considered if such releases are being proposed in different countries/regions. These include: details of the construct; details on the biology of the recipient, especially in relation to the ecology of the proposed release site; details of the proposed release site(s), especially considering sexually-compatible species in the agricultural and natural ecosystems; new potential non-target species; impact of new environment(s) on the phenotype of the GMO and possible production of factors that could affect food and/or feed (e.g., toxins, allergens or anti-nutrients); and any possible impacts on local agronomic systems.

The case studies:

Case Study 1. Herbicide tolerant Canola in Canada, raising the specific issue of the Canadian system with Plants with Novel Traits (PNTs) (including non-GM PNTs).

Case Study 2. Herbicide-tolerant plus insect-resistant corn in Argentina. The specific issue is the assessment of a GM crop with traits stacked by breeding where the individual traits have been previously assessed.

Case Study 3. Insect-resistant eggplant in India, with the production and assessment of this crop by a developing country.

Case Study 4. Virus-resistant papaya in Hawaii (USA) which was produced by a public organization and raises the issue of assessing a plant-induced pesticide which fell originally under the remit of the agency assessing non-GM crops.

Case Study 5. Slow ripening tomatoes. The issue here is the use of a transgene which does not produce a protein, but operates by affecting the normal plant metabolic pathway.

CASE STUDY 1

CANOLA MODIFIED TO GIVE MALE STERILITY AND HERBICIDE TOLERANCE

PREAMBLE

In this case study we examine three points: the general types of data in an application for release of a GM crop in many developed countries; the Canadian regulatory system and the differences with other regulatory systems; some more general aspects to be considered with the same trait in different crops. The application in this study was for a transgenic line with two (stacked) traits – stacked traits are discussed in Case Study 2.

I. THE PROPOSAL FOR RELEASE IN CANADA

Some of the major points in the proposal are outlined below; more details can be found in the references at the end of the case study.

A. Background

1. The Problem

As noted in Chapter 1, Box 1.5, F1 hybrid crops have enhanced vigour which gives improved growth and yield. The potential benefits of F1 hybrid canola are estimated to be a yield increase of 20–25% more than open-pollinated varieties, and uniformity which facilitates harvesting and marketing. One of the problems in producing F1 hybrids is to control the crossing between the two parents. This is often done by making the female parent sterile so that it will not self-fertilize and instead is fertilized with pollen from the other partner in the hybrid. However, this produces a male-sterile F1 for which, if seed is to be produced as the crop product, male fertility has to be restored. The development of hybridization systems for canola by traditional methods has not been completely successful for commercial applications. In this proposal both male sterility and fertility restoration are produced by transformation.

Weeds can cause significant yield losses in canola and can be difficult to control in conventional crops. By genetically modifying canola to be herbicide-tolerant, the target herbicide can be sprayed on the crop killing the weeds without affecting the crop.

2. The Specific Issues

As well as the food and feed safety aspects of the introduction and expression of the gene conferring herbicide tolerance, there are potential environmental risks of spread of the transgene to sexually compatible species. Canola is a promiscuous outcrosser and there is extensive transference of pollen between canola crops and to wild sexually compatible species. This could result in:

- The inadvertent transfer of the herbicide resistance trait to non-GM canola, potentially including canola grown organically and thus causing problems with organic crop regulations.
- Transfer of the herbicide resistance trait to canola varieties with other herbicide resistance traits, which could cause problems with the control of volunteer plants.
- Transfer of the herbicide resistance trait to sexually compatible species (e.g., *Brassica rapa*), which could cause a problem of herbicide-resistant weeds.

This proposal was for release in Canada where the regulations are directed at Plants with Novel Traits (PNTs) (see Chapter 6, Section I.C.1). This case study also shows how a herbicide-tolerant canola line produced using non-GM methods was assessed as a PNT under the Canadian system, and how this approach to GM F1 hybrid production was assessed in other regulatory systems.

B. Proposal

1. The Recipient

Canola (*Brassica napus*) (also known as oilseed rape or Argentinean rape) is an annual species grown for its oil production which is used in human food and livestock feed. In Canada, canola is spring sown as it has limited tolerance to cold. Weeds can be one of the limiting parameters to production, especially in the early stages of growth.

B. napus is an amphidiploid composed of two genomes, *B. rapa* and *B. oleracea*. It is out-pollinated and has been shown to outcross up to 30% with plants of the same species and potentially with related species (*B. rapa*, *B. juncea*, *B. carinata*, *B. nigra*, *Diplotaxis muralis*, *Raphanus raphanistrum* and *Erucastrum gallicum*). Gene flow is most likely to occur with *B. rapa* and other canola varieties, and only occasionally with the other species.

The recipient canola variety was cultivar Drakkar which is a low euricic acid, low glucosinolate (two toxic compounds originally bred to a low level by conventional breeding) spring variety.

2. The Donors

The male sterile line (MS8) was produced by the introduction of the barnase gene from a common soil bacterium, *Bacillus amyloliquifaciens*. The barnase gene encodes a ribonuclease (RNAse) that is expressed only in the tapetum cells of the pollen sac during anther development. This RNAse affects RNA production, disrupting normal cell function and arresting early anther development, thus leading to male sterility.

The restorer line (RF3) was produced by the introduction of the barstar gene, also isolated from *B. amyloliquifaciens*. The barstar gene codes for an RNAse inhibitor that is only expressed in the tapetum cells of the pollen sac during anther development and which specifically inhibits the barnase RNAse expressed by the MS8 line. Thus, when pollen from the restorer line RF3 is transferred to the male-sterile MS8 line, the resultant progeny express the RNAse inhibitor, allowing hybrid plants to develop normal anthers and restore fertility.

Both the MS8 and RF3 lines also contain the *bar* gene, isolated from the common soil microorganism *Streptomyces hygroscopicus*, which encodes a phosphinothricin acetyl transferase (PAT) enzyme. The active ingredient in phosphinothricin herbicides is glufosinate ammonium, which acts by inhibiting the plant enzyme glutamine synthesase leading to accumulation of phytotoxic levels of ammonia. PAT detoxifies glufosinate ammonia eliminating its herbicidal activity. The *bar* gene was introduced into both the MS8 and RF3 lines as a selectable marker to identify transformed plants during tissue culture regeneration, and as a field selection method to obtain 100% hybrid seed.

3. The Constructs

Details of the construct pTHW 107 used for transformation to produce MS8 and of construct pTHW 118 producing RF3 are shown in Figures 1A and 1B respectively; the figures show details of the constructs within the T-DNA borders.

The genetic components of the two constructs are listed in Table 1.

TABLE 1   Genetic elements in plasmids pTHW 107 and pTHW 118

Abbreviation	Element name	Source	Size (KB)
LB	Left border	T-DNA	–
P-TA29	Anther-specific promoter	*Nicotiana tabacum*	1.509
barnase[a]	Barnase	*Bacillus amyloliquefaciens*	0.446
barstar[b]	Barstar	*B. amyloliquefaciens*	0.3
T-nos	Nopaline synthase terminator	*Agrobacterium tumefaciens*	–
P-Ssu	Promoter for ribulose-1-5-biphosphate carbolase small subunit + 5′ coding sequence for chloroplast transit peptide	*Arabidopsis thaliana*	1.725
bar	Phosphinothricin acetyltransferase	*Streptomyces hygroscopicus*	0.55
T-g7	Gene 7 terminator	*A. tumefaciens*	0.211
RB	Right border	T-DNA	–

[a] Only in pTHW 107.
[b] Only in pTHW 118.

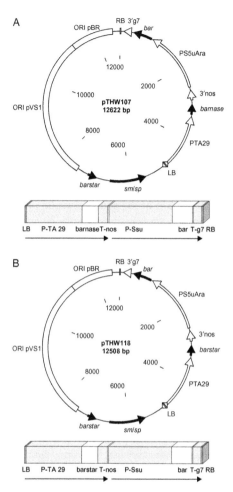

FIGURE 1 Details of the constructs used to make canola lines MS8 and RF3. For each, the complete plasmid is shown above and the inserted DNA is shown below.

The genetic elements of the construct were inserted between the left and right borders of the *Agrobacterium tumefaciens* T-DNA. The *barnase* and *barstar* genes were under the control of an anther-specific promoter (P-TA29) so that they were only expressed in the anthers. The *bar* gene was under control of the Ssu promoter to which was fused the 5'-terminal coding sequence of a chloroplast transit peptide to target it to chloroplasts.

Sequences outside the T-DNA region contained: colE1 replication region from *Escherichia coli*; pVS1 replication region from *Pseudomonas* sp.; and the streptomycin/spectinomycin resistance gene with its own promoter from *Klebsiella aerogenes*.

4. Transformation Method

Both MS8 and RF3 lines were produced using *Agrobacterium*-mediated transformation using a disarmed Ti plasmid (see Chapter 1, Box 1.7).

5. Analyses of Products

a. Southern Blot Analyses   Southern blot analysis of the genomic DNA from lines MS8 and RF3 showed that each contained a single site of insertion of the construct. In the MS8 line, the T-DNA was arranged in an inverted repeat structure with a second incomplete T-DNA copy which contained a functional part of the promoter P-TA29, the coding region for barstar, the 3′ *nos* terminator and a non-functional part of the promoter PSsu.

The insertion site of each line was characterized in detail. The insertion in MS8 was located in the *B. rapa* portion of the genome, while that of RF3 was in the *B. oleracea* portion of the genome.

Southern blots and PCR analyses showed that no sequences outside the T-DNA region of plasmids pTHW 107 and pTHW 118 were integrated into the plant genome. Thus, there were no antibiotic resistance marker genes in the transformed plants.

b. Expression and Mendelian Inheritance of Transgenes   Northern blots showed that the *bar* gene (PAT protein) was expressed in leaves and flower bud, but not in dried seed of both lines.

All the inserted genes of both lines were stable and inherited in a standard Mendelian manner for a single dominant locus.

c. Biology of MS8 and RF3 Lines   Important agronomic characters including germination, vegetative vigour and flowering period for both lines were tested under various environmental conditions, and were shown to be within the normal range of expression of these characteristics in unmodified *B. napus* counterparts. Similarly, characters such as time to maturity and seed production for RF3 and MS8 × RF3 hybrids were similarly within the normal range of expression.

Flowers of the MS8 line had undeveloped anthers, slightly smaller petals, and did not produce fertile pollen; however, nectar was normal and the plants were visited by pollinating insects. The barnase gene in line MS8 conferred male sterility which was restored by crossing in the barstar gene from line RF3. Thus, both these genes appeared to be functioning as expected.

d. Disease and Pest Characteristics of Transgenic Canola   The reaction of the transgenic lines to major *B. napus* pests and pathogens (e.g., flea beetles, diamondback moth larvae, blackleg, sclerotinia) fell within the ranges displayed by non-GM commercial varieties.

6. Environmental Consequences of the Introduction of Transgenic Canola

The two parental transgenic lines and the hybrid were field tested under confined conditions in Saskatchewan, Alberta, Manitoba and Ontario (the major canola growing provinces) over three years.

a. Outcrossing Potential with Other Species   The sterility of line MS8 would ensure that gene introgression from that line into cultivated or wild sexually-compatible plants would be extremely unlikely. MS8 plants could be pollen recipients, but the progeny would be partially male sterile.

Line RF3 and hybrid MS8 × RF3 displayed the normal reproductive characteristics of *B. napus* plants and did not differ in outcrossing ability from non-GM canola plants.

b. Weediness   Field studies demonstrated that lines MS8, RF3 and MS8 × RF3 did not differ from their non-GM counterparts in invasiveness and survival characteristics such as vegetative vigour and seed dormancy. Thus, the introduced male sterility, fertility restoration and glufosinate tolerance did not confer any potential for weediness.

Furthermore, the transgenic lines MS8, RF3 and the hybrid between them had no novel phenotypic characteristics which could extend their use beyond the current canola growing region. They did not differ significantly in their outcrossing ability or competitiveness from commercially-produced non-GM canola varieties.

c. Non-target Adverse Effects   The barnase and barstar proteins are only produced in the tapetal cell layer of anthers at a specific developmental stage, and their production does not result in altered toxicity or allergenicity properties. Field tests indicated that the GM lines MS8, RF3 and the hybrid MS8 × RF3 did not have significant adverse impact on organisms beneficial to agriculture or other non-target organisms.

7. *Food and Feed Considerations*

a. Dietary Exposure   Human consumption of canola products is limited to the refined oil, which contains virtually no protein. The introduced gene products were not detectable in the refined oil produced from the GM lines and from the hybrid between the two lines.

Animals are fed canola seed meal. The barnase and barstar proteins were not found in dried seed and the PAT protein was considered to be at levels too low to cause concern.

b. Nutritional Composition   The composition of refined canola oil from the MS8 × RF3 hybrid was compared to that of the refined oil from non-transgenic canola. Some statistical differences in fatty acid composition were found, but the fatty acids from the transgenic lines (including the erucic acid levels) were within the normal range of canola oil fatty acids. Oil from transgenic lines processed using protocols resembling industrial practices (e.g., cooking, pressing, desolventizing oil and meal, oil blending, oil refining and deodorizing) was equivalent to that from non-transgenic varieties.

c. Toxicity and Allergenicity   Amino acid sequences of barnase, barstar and PAT proteins showed no homologies with known protein toxins and allergens. The biophysical properties of these proteins were also examined, and no properties characteristic of toxins and allergens (e.g., resistance to heating or proteolysis) were found. Because only the processed oil from the transgenic lines and their hybrids is used for human consumption and it contained few or no proteins, there were no toxicity or allergenicity concerns for this product.

C. Risk Assessment and Determinations

1. Environment

The Biosafety Office of the Canadian Plant Health and Production Division which regulates the environmental aspects of release of PNTs considered the following points:

- Potential of the PNTs to become weeds of agriculture or to be invasive of natural habitats;
- Potential for gene flow to wild relatives whose hybrid offspring may become more weedy or more invasive;
- Potential for the PNTs to become plant pests;
- Potential impact of the PNTs or their gene products on non-target species, including humans;
- Potential impact on biodiversity.

Based on the data provided by the applicants, the Biosafety Office reached the following conclusions:

- Characteristics such as germination, vegetative vigour, flowering period, time to maturity, production of nectar and attractiveness to pollinating insects, and seed production were within the normal range of these characteristics in unmodified *B. napus* counterparts. The transgenic lines had no specific added genes for cold tolerance or winter survival.
- There was no change in susceptibility to pests and pathogens.
- The transgenic lines did not differ from unmodified canola in invasiveness of unmanaged habitats.
- The transgenic lines had no altered weed or invasiveness potential compared to currently commercialized canola varieties.
- A longer term concern is the potential development through gene flow of crop volunteers with resistance to multiple herbicides, which could result in the loss of use of these herbicides.
- Gene flow from the transgenic lines to canola relatives is possible, but would not result in increased weediness or invasiveness of these relatives.
- The plant pest potential of these lines had not been inadvertently altered.
- The unconfined release of these transgenic lines would not result in altered impacts on interacting organisms, including humans, compared with currently commercialized non-transgenic counterparts.
- The potential impact on biodiversity of these transgenic lines was equivalent to that of currently commercialized non-transgenic canola varieties.

Based on these conclusions, the unconfined release to the environment and feed use (see below) of MS8, RF3 and MS8 × RF3 was authorized. Any other *B.*

napus lines and intra-specific hybrids resulting from the same transformation events, and all their descendents, could also be released provided no inter-specific crosses were performed, provided the intended use was similar, provided it is known that these plants do not display any additional novel traits and provided the resulting lines can be shown to be substantially equivalent to currently grown canola in terms of potential environmental impact and livestock feed safety.

2. Food and Feed

Food safety of PNTs is regulated by the Health Protection Branch of the Food Directorate, Health Canada who considered the information supplied by the submitter. They concluded that, as human consumption of canola products is limited to the refined oil which contains little or no proteinaceous material, there would be no dietary exposure to the proteins introduced into the transformed lines or their hybrids. Of special note was the fact that the transformation process had not altered the levels of potential toxins, erucic acid and glucosinolate. The nutritional composition of the refined canola oil from the transformed lines and the hybrids was within the normal range for canola oil fatty acids and, after processing by industrial practices, had compositional and physical characteristics equivalent to non-GM canola.

Based on these factors, the food use of refined oil from these PNTs was authorized. The regulation of feed safety falls under the Biosafety Office of the Canadian Plant Health and Production Division (which also regulates environmental safety – see above). They considered the information on anti-nutritional factors and nutritional composition of the transformed lines and the hybrids provided by the applicant, and concluded that these parameters were substantially equivalent to untransformed canola varieties.

II. OTHER RELATED SITUATIONS AND ISSUES

A. Other Regulatory Authorities

Canola is grown in many temperate countries, and canola oil and oil seed feed are used even more widely for human food and animal feed, respectively. Thus, approval for its use in countries other than Canada falls under other regulatory authorities.

1. The European Union

An application was made by the competent authority of Belgium for placing *B. napus* lines MS8, RF3 and MS8 × RF3 (described above) on the market (for the import, cultivation and use as an animal feed, but not for human food). According to the EU procedures, the Belgian competent authority prepared an assessment report which was considered by the commission and the competent authorities of other member states. The competent authorities of certain member states raised objections to the placing of these products on the market on points that included potential allergenicity and toxicity, possible effects of pollen flow and spillage, lack of a monitoring plan, lack of labelling proposals and possible problems in the detection of the products (e.g., distinguishing the hybrid from a mixture of the two parental lines). The European Food Safety Authority (EFSA) gave the opinion that these products were as safe as conventional canola (oilseed rape) for humans and animals, and in the context of intended uses, for the environment. The application was modified, but there were still objections by the member states in the regulatory committee and at council level, neither of which could reach a qualified majority decision. In the light of the inability to reach qualified majority decisions and of the opinion of EFSA, the European Commission decided that these products could be imported and processed for use as animal feed, but could not be cultivated. They also decided that management systems should be put in place to prevent these lines entering cultivation. This involves unique identifiers for the GM products and monitoring.

In contrast to Canada, imidazolinone-tolerant canola (produced using mutagenesis by another company to the line described in Section II.B) is not regulated in Europe, as the EU does not especially regulate such products. Imidazolinone-tolerant canola comprised 20% of the canola planted in Europe in 2000 and 2001.

2. Australia

The Australian competent authorities considered an application for the unconfined release of *B. napus* lines MS8, RF3 and MS8 × RF3 for growing, food and feed. The risk assessment followed similar lines to that described for Canada in Sections I.B and I.C. There were controlled field releases to examine ecological impacts (including indigenous non-target organisms) and food and feed properties for the lines grown under Australian conditions. The data from these trials did not differ significantly from the Canadian contained field trials.

The Australian competent authority concluded that the GM lines did not differ significantly from conventional canola. However, release considerations also involved governmental agencies other than the GM competent authority and the licence for uncontained release had various conditions:

- A herbicide resistance management plan was required by the authority which regulates the use of herbicides;
- An industry stewardship proposal focusing on good agricultural practices and good handling was required. This plan would incorporate:
  - separation of the GM and non-GM crops to the extent required by the market. This would be to protect the non-GM crop market and the export market
  - maximization of the effective use of the technology
  - contribution to agricultural sustainability.
- The releasing company had to report to the regulatory authority on the amount of seed sold commercially and otherwise grown in each growing season for each state and territory.

As with the EU, imidazolinone-tolerant canola (produced using mutagenesis by another company to the line described in Section II.A) is not regulated in Australia.

B. Comparable PNTs in Canada

The Canadian regulations cover Plants with Novel Traits (PNTs) which can also be produced by techniques other than genetic manipulation. Below is an example of an application to release a non-GM PNT.

1. The Proposal

As noted in the proposal above, weeds can be a serious problem in canola crops. The application was for the release of three canola (*B. napus*) lines (NS738, NS1471 and NS1473) tolerant to herbicides based on imidazolinone. Imidazoline herbicides are active against the plant enzyme acetolactate synthase (ALS), also known as acetohydroxyacid synthase or acetolactate pyruvate-lyase, which catalyses the first step in the biosynthesis of the essential branched chain amino acids isoleucine, leucine and valine.

The modified lines were developed from the *in vitro* culture of microspores from *B. napus* cultivar Topas which was mutagenized with ethyl nitrosourea and selected on a culture medium containing imidazolinone. The mutagenesis modified the ALS gene sufficiently to alter the binding site to imidazolinone, such that the herbicide no longer inactivated the ALS enzyme. The tolerant regenerated haploid plantlets were recovered and treated with colchicine to induce chromosome duplication resulting in double-haploid plants. Two imidazolinone-tolerant lines were selected, crossed reciprocally, then crossed with *B. napus* cultivars Topas and Regent, and the progeny subjected to repeated cycles of selfing and further breeding.

2. Product Data for the Regulatory Authority

The types of data submitted on the modified lines were similar to those described above for the application for release of GM canola. The data were obtained from growing the lines at 21 locations representing four different growing environments. The agronomic and environmental safety analysis showed that seed production, vegetative vigour, days to flowering, time to maturity, plant lodging, plant height, production of oil and protein, disease susceptibility and potential for weediness were within the range of unmodified *B. napus* counterparts. The levels of valine, isoleucine and leucine were similar to those of unmodified canola cultivars, showing that the ALS activity of the PNT was not affected by the mutation. Analyses of amino acid composition and of anti-nutritional factors (erucic acid and glucosinolates) showed substantial equivalence to the unmodified counterparts.

3. Risk Assessment and Determination

The risk assessment on these non-GM lines was similar to that described above for the GM canola. The determination was that these modified lines were substantially equivalent to unmodified canola lines, and thus could be released to the environment and used for food and feed in an unconfined manner. However, there was a longer-term concern expressed on the potential for gene flow producing canola with resistance to several different herbicides which would cause problems with crop volunteers.

C. Other Issues

1. Herbicide-resistant and Male Sterile Canola

The above examples show the basic approaches to risk assessment for the release of specific GM and non-GM canola lines in three different regulatory systems. The approach to environmental and food and feed safety assessment is basically the same, but the decisions reached depend on the regulatory system. The assessment of the release of, for example, GM canola in different countries would depend on several local factors including:

- The agronomic system in that country – is there increased potential for weediness or invasiveness of the GM crop or of any weed species that are sexually compatible?
- Are there non-target organisms specific to the country in question which could be affected?
- Do the local growing conditions affect the food and feed properties of the crop product (e.g., toxicity, allergenicity, food composition)?
- The cropping systems in the country – would gene flow to sexually-compatible crops have any impact on local markets (e.g., organic products) or international trade?
- National opinion about GM products.

2. The Biology of Canola

As noted in Section I.A.2, canola is a promiscuous outcrosser and is sexually compatible with several other species. It provides a good example of the need to have a detailed understanding of the crop species being assessed and the ecology of the environment into which it is being released. For example, the sexually-compatible *Brassica rapa* and *B nigra* are grown as crops in temperate countries, and so there is the potential for GM traits in canola to be passed onto them. These species, and *B. juncea*, are classed as potential weeds in the US, and therefore the acquisition of herbicide tolerance could pose problems. *Rhaphanus raphanistrum*, also sexually-compatible with canola, is native to Europe, North Africa and South West Asia, and has been introduced elsewhere. It is becoming a problem in some parts of Australia, where it has become naturally tolerant to some herbicides. It should also be remembered that pollen flow between a crop and a sexually-compatible wild species can be a two-way process, and consideration should be given to a wild species being a source of a GM trait into a non-GM crop.

Reference Sources and Further Information

http://www.agbios.com/static/cropdb/LONG_MS8xRF3_printer.html and links therein.
http://www.inspection.gc.ca/english/plaveg/pbo/dd9617e.shtml.
http://www.inspection.gc.ca/english/plaveg/bio/dd/dd9503e.shtml.
http://www.hc-sc.gc.ca/fn-an/gmf-agm/appro/28bg_pgs-eng.php.
http://www.europabio.org/InfoOperators/MS8RF3/M8RF3%20Brochure.pdf and links therein.
http://www.biosafety.be/GMCROPFF/EN/TP/SBB_NotificationC_BE_96_01.html.

CASE STUDY 2

CORN MODIFIED TO BE RESISTANT TO LEPIDOPTERAN PESTS AND TOLERANT TO GLYPHOSATE HERBICIDE

PREAMBLE

This case study looks at two issues. First, it examines a proposal to release a GM corn line into Argentina, originally developed in the US, which contained two transgenic traits that had been previously assessed and approved in Argentina and were combined by conventional breeding techniques. The creation of this stacked corn line by conventional breeding raises the following question:

• Should transgenic lines with stacked traits be assessed solely based on the two individual traits, should they be newly assessed as a novel line, or should they be assessed as some combination of these two approaches?

Secondly, the Case Study discusses the potential problems that regulatory authorities may face in dealing with stacked traits more generally.

I. THE PROPOSAL FOR RELEASE IN ARGENTINA

Some of the major points in the proposal are outlined below; more details can be found in the references at the end of the case study.

A. Background

1. The Problem

Argentina and Brazil are the second biggest exporters of corn products after the US. Most of the corn is for animal feed, corn oil and increasingly for biofuels.

Both lepidopteran pests, especially the sugarcane borer, *Diatraea saccharalis*, and weeds are important to corn cultivation in Argentina. Transgenic lines expressing the Bt toxin Cry1Ab (see Box 4.3 for Bt toxins) which confers resistance to stalk borers like *Diatraea* spp. and expressing 5-enolpyruvylshikimate-3-phosphate (CP4 EPSPS) giving tolerance to the glyphosate herbicide were separately produced. These lines were then crossed by conventional breeding techniques to produce a line containing both transgenic traits (stacked traits).

2. GM Regulation in Argentina

Argentina's legislative framework for regulating GM organisms, established in 1991, follows a process-based approach (see Chapter 6, Section I.C.2). Regulatory oversight is the responsibility of The National Committee on Agricultural Biotechnology (CONABIA) (which is the lead agency); The National Service of Agrifood Health and Quality (SENASA); and The National Institute of Seeds (ex-INASE). The National Directorate of Agricultural Food Markets (DNMA) also assesses the potential impact that commercialization of a GM crop might have on Argentina's export markets.

Authorization for the commercialization of a GM crop in Argentina is a 3-step process which normally takes about two years.

- The first step, after CONABIA has made an initial biosafety assessment, is "flexibilization" of testing conditions, which involves authorization for unconfined field trials.
- Secondly, SENASA evaluates the food and feed safety of the GM crop.
- In the third step, DMNA assesses the impact of the GM crop on the export market security.

A flow chart of this process is shown in Fig. 1.

After passing these three steps, CONABIA prepares a recommendation which is submitted to the Agricultural Directorate of the Secretariat of Agriculture, Livestock, Fisheries and Food (SAGPyA), which authorises commercialization. The applicant then needs to apply for seed registration and, if the crop contains pesticide substances, a pesticide registration (as for conventionally-bred plants).

B. Proposal

The two parental lines (glyphosate-tolerant and insect resistant, respectively) were produced independently and had previously been separately approved for release in Argentina, where they are deemed to be safe. The questions that might need to be addressed for the breeding stack were:

- Does the combination of the characters of the two individual parents give rise to any new risks not identified in the individual parent lines?
- Does the bringing together of the two transgenic genotypes by conventional breeding create any new risk not present in the parents?

The data provided to the Argentinean regulatory authority were a combination of the original data for the separate events, data generated in the US on the breeding stack, and locally generated data on the stack. Summarized below are some of the primary pieces of information that were provided.

1. The Parents

The information on the recipient species, the donors of the inserted sequences, the constructs and the transformation method for the two parents were given in the applications for their individual releases.

2. Analyses of Product (the Breeding Stack)

Full analyses on the parental lines had been undertaken to gain approval for their separate releases in Argentina and in other countries such as the US; the types of data provided for these release applications were similar to those shown in Case Study 1. For the US, where this stack was first released commercially, there was no requirement to undertake a separate full analysis of the breeding stack, as the Environment Protection Agency (EPA) does not regulate non-pesticidal traits such as herbicide tolerance. However, for Argentina (and other countries where such a stack is viewed differently to the US), bridging data are

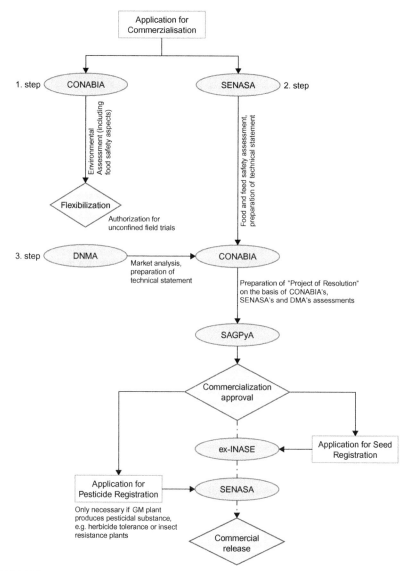

FIGURE 1   Steps to commercialization of genetically modified crops in Argentina. From ISNAR (2002).

generated on breeding stacked traits including molecular characterization of the stack (to compare to the parental lines), expression, and interaction studies to verify that the two traits do not produce synergistic or antagonistic pesticidal effects.

In this case, an analysis indicated that there would be no interactions among the gene products or negative synergistic effects in the stacked product. The Cry1Ab protein is not an enzyme, and therefore would not directly affect plant metabolism. The CP4 EPSPS has a high affinity for its substrates (phosphoenol-pyruvate and shikimate-3-phosphate) and would not be expected to interact with the Cry1Ab.

Argentinean regulatory authorities reviewed the data on the stacked herbicide-tolerant insect-resistant product, including molecular characterization, expression and composition, even though the individual events were already approved in Argentina.

C. Risk Assessment and Determinations

The individual parental lines of this stacked hybrid had been assessed and approved for release in Argentina previously, the insect-resistant line in 1998 and the herbicide-tolerant line in 2004. Those submissions included both US data and locally-obtained data. The Argentinean authorities then reviewed the information on the stacked package which included additional locally-obtained data and data from the US.

As noted in Section I.A.2, the Argentinean regulatory system is a three-step process, with the third step considering the impact that a release might have on the export market security. At the time the release of this stacked hybrid was being considered, the herbicide-tolerant parent, and therefore the breeding stack, had not been approved by the European Union, an important export market. However, it was considered that this would not be a complete impediment to the release of the stacked product in Argentina.

II. OTHER RELATED SITUATIONS AND ISSUES

A. *Types of Stacked Products*

There are three situations in which transgenic traits can be stacked in a single crop line:

- First, the two (or more) genes can be introduced at the same time to give what might be termed "co-transformational or vector stacking." This is considered in Case Study 1.
- Secondly, transgenes can be brought together by conventional breeding ("conventional breeding stacking"), as is described in this case study.
- Thirdly, there can be "inadvertent stacking," due to uncontrolled gene flow between two released lines, each containing a single transgene.

Co-transformational (vector) stacking is deemed by all regulatory authorities as a new event, and approval for release of such lines would follow the same risk assessment and decision-making process as a single-trait transformant.

Regulatory authorities differ in their approach to conventional breeding stacks. As described in this case study, the US authorities treated this stacked product as a conventionally-bred line which does not go through the GM line approval procedure. Regulatory authorities in Canada and Australia pursue a similar approach, requiring some sort of notification for such stacks. Other authorities, e.g., Argentina, Japan and the EU, require more data on a breeding stack of GM traits, varying from a bridging approach to viewing a stack as an entirely new event. Even the US authorities differentiate between breeding stacks depending on the traits involved; where stacks are of PIP traits (see Case Study 4) bridging data similar to that required by Argentina are required by the EPA.

Inadvertent stacking (sometimes called field stacks) is likely to be an increasing situation in the future. At least some regulatory authorities are recognizing the possible situation when granting approval for the release of single transgene lines for some traits, and ask for various management schemes. The potential problem will vary according to the trait in question, the biology of the crop species and the scale of planting. The stacking of different herbicide tolerances could cause difficulties in the control of sexually-compatible weeds and of volunteers in outcrossing crop species such as canola (see Case Study 1). It has less potential for leading to problems with poorly outcrossing crops such as corn or soybean, and with vegetatively propagated crops such as potato.

In the future, the stacking of more complex traits with less familiarity than the herbicide tolerance and insect resistance described in this case study, such as combining traits targeted at a biotic factor (e.g., insects), an abiotic factor (e.g., drought tolerance) or a food quality factor (e.g., nutritional enhancement) may raise other issues. These include potential interactions among the traits and potential environmental impacts, and are likely to be treated on a case-by-case basis until much more familiarity is gained.

Reference Sources and Further Information

http://www.agbios.com/static/cropdb/LONG_NK603_x_MON810_printer.html.

Burachik, M., and Traynor, P.L. (2002). Analysis of a National Biosafety System. Regulatory Policies and Procedures in Argentina. International Service for National Agricultural Research (ISNAR) Report 63, April.

De Schrijver, A., Devos, Y., van den Bulcke, M., Cadot, P., de Loose, M., Reheul, D. and Sneyers, M. (2007). Risk assessment of GM stacked events obtained from crosses between GM events. *Trends Food Sci. Technol.* **18**, 101–109.

CASE STUDY 3

INSECT-RESISTANT EGGPLANT IN INDIA

PREAMBLE

Case studies 1 and 2 focused on risk assessment of transgenic lines produced in developing countries and assessed by regulatory authorities in those countries; in Case Study 2 the transgenic crop was subsequently introduced to, and assessed in, a second country. In Case Study 3 we examine a transgenic line produced in a developing country and assessed by the regulatory authority in that country.

I. THE PROPOSAL

A. Background

Some of the major points in the proposal are outlined below; more details can be found in the references given at the end of this case study.

1. The Problem

Eggplant (brinjal) has been cultivated in India for 4000 years, and is the principal solanaceous vegetable crop grown in most parts of that country. It is mainly cultivated on small family farms and is a source of cash income for resource-poor farmers. This staple vegetable crop is extensively damaged by the lepidopteran insect, the brinjal fruit and shoot borer (*Leucinodes orbonalis*) which can cause losses up to 70%. The pest poses a serious problem because of its high reproductive potential and rapid turnover of generations and the intensive cultivation of brinjal in both the wet and dry seasons. The current control is by the use of insecticides with 25–80 spray applications per crop (average 4.6 kg active ingredient per hectare) leading to substantial build-up of pesticide residues in the produce, destruction of beneficial insects (non-target organisms), farm workers being exposed to pesticides, and environmental pollution.

The use of plants transformed to express *Bacillus thuringiensis* (Bt) delta-endotoxin proteins (see Chapter 4, Box 4.3), including Cry1Ac and other Cry1 and Cry2 proteins, has been shown to be very effective at controlling lepidopteran insect pests of several crops. Several Bt cotton products designed to control cotton bollworms have been approved in India.

The case study under consideration here is on a transgenic Bt eggplant (brinjal) line [Elite Event-1 (event EE-1)] produced in India and assessed using studies done in India.

2. The Specific Issues

Brinjal grown in India is produced mainly for human consumption. Thus, the risk assessment considers both food (and feed) safety, and the potential environmental impact of the introduction.

B. Proposal

1. The Recipient

The recipient was *Solanum melongena*, which belongs to the family Solanaceae. There are three main varieties of *S. melongena*; round or egg-shaped cultivars grouped under var. *esculentum*, long, slender types included under var. *serpentinum* and dwarf plants grouped under the var. *depressum*. There are four *Solanum* species related to *S. melongena*: *S. incarnum*; *S. xanthocarpum*; *S. indicum*; and *S. maccani*. *S. melongena* is readily crossable with *S. incarnum*.

The centre of origin of *S. melongena* is unknown, but is possibly India or Southern China.

S. melongena is usually self-pollinated, but in some circumstances the extent of cross-pollination has been reported to be as high as 48%. Where outcrossing occurs, the plants are primarily insect-pollinated.

Brinjal is a perennial herbaceous plant which is grown commercially as an annual crop.

2. The Donor

The Bt *cry1Ac* toxin gene was obtained from *B. thuringiensis* subsp. *kurstaki* HD73. This toxin damages the midgut of a narrow range of lepidopteran insect species, leading to their death. It does not harm non-lepidopteran insects, other invertebrates, mammals, fish or birds which lack appropriate toxin receptor sites in their midgut.

The *nptII* antibiotic resistance gene, used as a selective marker to identify transformed plants during tissue culture regeneration and multiplication, was derived from a bacterial transposon (Tn5 from *Escherichia coli*). This gene product inactivates aminoglycoside antibiotics such as kanamycin and neomycin.

The *aad* gene which encodes for the bacterial selectable marker enzyme 3"(9)-O- aminnoglycoside adenyl transferase (AAD) allowed for the selection of bacteria containing the pMON 10518 plasmid on media containing spectinomycin or streptomycin. The *aad* gene is under the control of a bacterial promoter, and hence is not expressed in Bt brinjal. The *aad* gene was isolated from transposons *Tn7*.

3. The Construct

The plasmid vector pMON 10518 (Fig. 1) which contains the *cry1Ac* gene expression cassette, together with the *nptII* and *aad* genes, was transferred to *Agrobacterium tumefaciens* strain LBA4404.

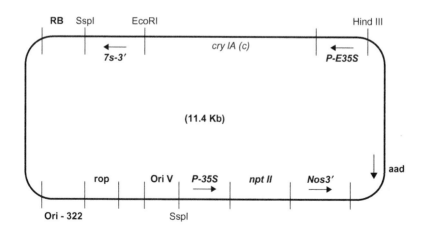

FIGURE 1 Plasmid for transforming brinjal with the Bt gene. The significant genes are defined in Table 1; Ori-322, rop and Ori V are sequences needed for the bacterial replication of the plasmid. RB is the T-DNA right border; EcoR1, Hind III and Ssp1 are restriction endonuclease sites.

Table 1 shows details of the components of the inserted construct.

TABLE 1   Details of the genetic construct introduced into EE-1 brinjal

Abbreviation	Element name	Source	Size (kbp)
P-E35S	Double-enhanced 35S promoter	*Cauliflower mosaic virus*	0.62
Cry1Ac	Cry1Ac delta endotoxin gene	*Bacillus thuringiensis* ssp. *kurstaki*	3.5
7s-3'	3' non-translated terminator sequence	Soybean α subunit of the β conglycinin gene	0.43
P-35S	35S promoter	*Cauliflower mosaic virus*	0.32
nptII	Neomycin phos-photransferase II gene	*Escherichia coli*	0.79
nos	3' non-translated terminator	*Agrobacterium tumefaciens*	0.26
aad	3″(9)-O-aminoglycoside adenyl transferase	*Escherichia coli* transposons Tn7 with bacterial promoter and terminator	0.79

4. Transformation Method

Young cotyledons of eggplant were transformed using the Agro-bacterium-mediated method (see Chapter 1, Section III.C.2). Transgenic plants were regenerated through tissue culture procedures on media containing kanamycin as a selection marker. Selected plants were analysed using the enzyme-linked immunosorbent assay (ELISA) to test for the presence of Cry1Ac protein.

The chosen line was termed event EE-1.

5. Analyses of Products

a. Southern Blot Analyses   Southern blot analyses showed that a single copy each of the Bt *cry1Ac* and the *nptII* genes were inserted into the selected line.

b. Expression and Mendelian Inheritance of Transgenes   The Bt transgene in the transgenic Bt brinjal behaves as a single gene, dominant Mendelian factor, and is stably integrated in the plant genome.

c. Biology of EE-1 Line   EE-1 line brinjal plants had similar germination, growth characteristics, pollen production and boll production when compared with untransformed parent plants.

d. Expression of Cry1Ac in Transgenic Brinjal   The concentrations of in-planta expressed Bt insecticidal protein, Cry1Ac, in various tissues (leaf, shoot, stem, flower, fruit and root) were quantified using a quantitative ELISA with tissues from non-Bt plants as controls. Cry1Ac was not detected in any of the non-Bt samples. The levels of Cry1Ac protein varied from 5 to 47 ppm in shoots and fruits of EE-1. These levels of Cry1Ac protein tissues compare well with the concentration needed to effectively control *Leucinoides orbonalis* larvae, which was calculated to be 0.059 ppm Cry1Ac.

e. Disease and Pest Characteristics of Transgenic Brinjal The Cry1Ac protein produced in EE-1 brinjal is comparable in activity to naturally occurring Cry1Ac protein. NPTII and AAD proteins are used as selectable markers, have no pesticidal activity and are not known to be toxic to any non-target species.

6 *Environmental Consequences of the Introduction of Transgenic Eggplant*
a. Pollen Flow Pollen flow studies on Bt brinjal were conducted at two different locations. At each location a central block containing Bt brinjal was surrounded by concentric rings of non-Bt brinjal to assess the distance travelled by the transgene and the outcrossing percentage. The maximum distance that the pollen travelled was 20 metres, with outcrossing potential varying between 1.6% and 2.7%.

b. Germination and Weediness To assess the weediness of Bt brinjal, the rate of germination and vigour were compared, in both laboratory and field tests, to the non-transformed counterpart. There were no substantial differences between Bt and non-Bt brinjal in germination and vigour, indicating that there is no significant difference between transgenic Bt and control non-Bt brinjal with regard to weediness potential. In a further field study to monitor weediness potential, the area which had been planted with Bt brinjal and then harvested was left undisturbed and irrigated on a regular basis to allow for germination of any seeds that might have remained in the ground after harvesting the main crop (the plot was observed up to three months after final harvesting). No brinjal plants were observed to germinate or grow in this plot for the period of the study, which suggests that EE-1 does not have weedy characteristics. Thus, Bt brinjal does not exhibit any different agronomic or morphological traits compared to non-Bt brinjal/controls that would increase its potential to become a weed.

c. Possible Effects on the Soil To assess possible risks associated with accumulation and persistence of the plant-produced Bt proteins in soil where Bt brinjal is repeatedly grown, soil samples were collected from selected locations over a period of three years. Cry1Ac levels in the soil were measured, and impacts on soil microflora and soil invertebrates were studied. There were no differences between Bt and non-Bt plots with respect to soil bacteria and fungal counts both at the rhizosphere and the soil beyond the rhizosphere. Similarly, no significant variation was observed in numbers of important soil invertebrates such as earthworms and Collembola.

d. Impact on Non-target Organisms Studies at 17 locations showed that beneficial insects such as chrysopa, lady beetles and spiders were active in both Bt and unsprayed non-Bt brinjal crops.

7. *Food and Feed Considerations*
a. Substantial Equivalence of Bt Brinjal The chemical composition of the fruits, leaves, stems and root tissues of Bt brinjal were compared with similar samples from

three non-Bt controls. There were no statistical differences between Bt and non-Bt brinjal in protein, carbohydrate, oil, calories, ash, nitrogen, crude fibres, moisture content, calories in fruit tissue and nitrogen content in leaf, stem and root tissues.

b. Cooking Studies Cooked brinjal fruits are consumed in various forms in India. Tender Bt brinjal fruits were used in studies to determine whether the Bt protein was present in the cooked fruits. The Bt protein was undetectable in the cooked fruits at the first sampling time-point, irrespective of the cooking method used (roasted, shallow-fried, deep-fried or steamed). The first sampling time-point was five minutes for roasted fruit and one minute for the other forms of cooking. This study indicates that the Cry1Ac protein in Bt brinjal fruits is rapidly degraded on cooking.

C. Toxicity and Allergenicity

Acute oral administration of transgenic Bt brinjal expressing Cry1Ac protein to Sprague–Dawley rats at the limit dose of 5000 mg/kg did not cause any toxicity. Proteins that are non-toxic by the oral route are not expected to be toxic by dermal or pulmonary routes. The no-observed-adverse-effect-level (NOAEL) of transgenic Bt brinjal expressing Cry1Ac protein in Sprague–Dawley rats following oral administration for 90 days was found to be more than 1000 mg/kg body weight. This study demonstrates that Bt brinjal expressing Cry1Ac protein is non-toxic to the study animal by oral route.

Potential allergenicity of Bt brinjal was assessed using three tests:

1. The relative allergenicity of transgenic Bt brinjal was compared to that of conventional brinjal (non-transgenic), as measured by active cutaneous anaphylaxis (ACA) in Brown Norway Rats sensitized with brinjal. Six- to seven-week-old Brown Norway Rats were randomly selected and used for the studies. The animals were observed daily for signs of toxicity and pre-terminal deaths, weekly body weights and food consumption. There were no clinical signs of toxicity and pre-terminal death (mortalities). The weekly mean body weights increased in all the groups. There was no statistically significant intergroup difference in body weights between treatment and control groups. There were no significant differences in food consumption between treatment and control groups. There were no differences among the skin reactions of each of four extracts on the same animals. These observations suggest that there are no differences between the allergenicity or inflammatory characteristics of the brinjal extracts tested including transgenic Bt brinjal and non-transgenic brinjal. Statistical analysis of this study concluded that there are no biological differences between the allergenicity response among all the brinjal hybrids, including transgenic Bt brinjal and non-transgenic brinjal.
2. Transgenic Bt brinjal expressing Cry1Ac protein applied to intact rabbit skin for four hours did not cause any skin reaction throughout the observation period. The irritancy index was also 0.0. The observations and results of this study indicate that transgenic Bt brinjal expressing Cry1Ac protein can be classified as non-irritant to skin in rabbit.

3. Application of transgenic Bt brinjal expressing Cry1Ac protein to the vaginal mucous membrane of female rabbits did not cause any erythema or edema as observed for 72 hours after application. Based on the average irritation index (0.0), transgenic Bt brinjal expressing Cry1Ac protein was classified as non-irritant to mucous membrane in rabbit.

d. Alkaloid Content Comparison of Bt and Non-Bt Brinjal Extracts of fruits and roots of Bt and non-Bt brinjal were chromatographed over silica gel and eluted to obtain two alkaloids, namely Solamargine (molecular weight 867) and Solasonine (molecular weight 883). The structure of alkaloids was identified based on extensive 1-D and 2-D NMR and other spectroscopic studies. The alkaloid profile from Bt and non-Bt brinjal were the same, with no appreciable variation in their relative abundances.

e. Nutritional Studies Food products derived from brinjal are extensively processed before use for human consumption. Therefore, no intact protein or genetic materials are expected to be contained in food products derived from brinjal. Brinjal has a history of safe use as a source of food in India. Brinjal is a highly productive crop, and the fruit are consumed as cooked vegetables in various ways. Brinjal is a good source of minerals and vitamins, and rich in total water soluble sugars, free reducing sugars and amide proteins, among other nutrients. The fundamental principle of substantial equivalence when applied to Bt brinjal and its non-Bt counterpart revealed that Bt brinjal is substantially equivalent in its composition to control brinjal, and thus the food and feed derived from Bt brinjal will also be substantially equivalent to food and feed derived from its non-Bt counterpart. In addition to compositional analysis, the wholesomeness of feed from Bt brinjal was demonstrated in separate feeding studies with fish, chickens, cows, goats and rabbits.

8. Resistance Management Strategies for Bt Brinjal

As discussed in Chapter 4, Section V, it is considered prudent to adopt management strategies to minimize the risk of resistance developing in target insect species after release of insect-resistant crops. To address the possible strategies that could be employed to reduce the likelihood of target insects developing resistance to the Cry1Ac protein in India, computer simulations and laboratory and field studies were conducted to evaluate strategies for managing caterpillar resistance to the Cry1Ac protein. The results from these experiments, combined with an understanding of brinjal production and agronomic practices, provide the basis for a sound, practical, resistance management plan. The following have been identified as key components of the resistance management plan for Bt brinjal in India: monitoring for baseline susceptibility; resistance monitoring; assessment of level of control; refuge design and placement; remedial action plan; encourage integrated pest management (IPM) and farmer field days and educational programmes.

The concept of refugia is discussed in Box 4.11. Computer modelling suggested that the most effective refugia for Bt brinjal deployment would be 5% of the field to be planted to non-Bt brinjal. Farmers would be advised to plant the non-Bt refugia on two sides of the Bt crop.

C. Risk Assessment and Determinations

1. The Indian Regulatory System

The two main agencies responsible for implementation of the biosafety legislation (rules) are the Ministry of Environment and Forests (MoEF) and the Department of Biotechnology (DBT). The biosafety legislation also defines competent authorities and the composition of such authorities for handling various aspects of the rules. There are six competent authorities:

- Recombinant DNA Advisory Committee (RDAC);
- Review Committee on Genetic Manipulation (RCGM);
- Genetic Engineering Approval Committee (GEAC);
- Institutional Biosafety Committees (IBSC);
- State Biosafety Coordination Committees (SBCC);
- District Level Committees (DLC).

Out of these, the three agencies involved in approval of new transgenic crops are: IBSC set-up at each institution for monitoring institute level research in GMOs; RCGM functioning in the DBT to monitor ongoing research activities in GMOs and small-scale field trials; and GEAC functioning in the MoEF to authorize large-scale trials and environmental release of GMOs.

The route which an application for GMO release follows is shown in Fig. 2.

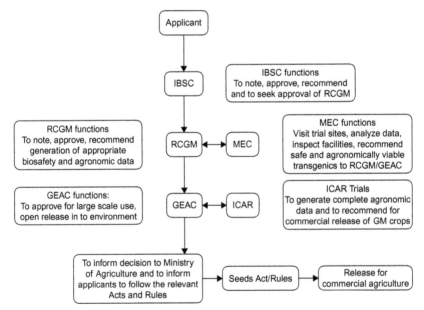

FIGURE 2 The route which an application for GMO release in India follows. Abbreviations as in text, plus MEC = Monitoring and Evaluation Committee. http://www. agbios.com/docroot/articles/05-265-003.pdf

2. Determination on EE-1

a. The application was for the experimental seed production of seven brinjal hybrids between EE-1 and non-GM brinjal varieties. The conditions were that the release should be within an institutional research farm, and that it should be strictly supervised by the Director of Horticultural Research or the Director of Research of the State Agricultural University with a view to facilitate the monitoring and supervision mechanism as stipulated by GEAC.

b. A subsequent application for full commercial release is going through regulatory review at this time.

II. OTHER ISSUES

A. Need for Monitoring

The monitoring of the release of EE-1 and other Bt Cry1Ac lines in India has identified two potential problems:

- Variation of Bt production was detected in some hybrids and also variation with the age of the plant.
- There was a problem of build-up of other cotton pests (mainly sap-sucking insects) due to the reduction of spray applications (reduced from about 30 to about 5 per crop).

These problems are not restricted to Indian conditions; such secondary pest problems related to changes in insecticide use have also been observed in Bt cotton in China.

B. Other Regulatory Authorities

As noted in Section II. C below, it is likely that Bt brinjal will be released in other countries. For such releases, although much of the basic risk assessment will have been done in India, it may be necessary to address some local issues such as differences in climate and ecology.

C. Public/private Sector Partnership

Although not strictly risk assessment and management, this initiative is a good example of the advantages of partnership between the public and private sectors. Bt brinjal was produced in the private sector, who donated the technology to public institutions in India, Bangladesh and the Philippines for use in open-pollinated brinjal varieties in order to meet the specific needs of resource-poor farmers. The sharing of knowledge and experience of the regulatory process for Bt brinjal in India could greatly simplify and lighten the regulatory burden in Bangladesh and the Philippines by eliminating duplication of the assembly of much of the risk assessment data. This, in turn, could contribute towards harmonization of regulations over the region.

Sources of Information

http://www.envfor.nic.in/divisions/csurv/geac/information_brinjal.htm.
http://www.envfor.nic.in/divisions/csurv/geac/decisions-june-85.pdf.
The Indian GMO Research Information System (IGMORIS) (http://igmoris.nia.in).
Choudhary, B. and Gaur, K. (2009). The development and regulation of Bt brinjal in India (eggplant/aubergine). International Service for the Acquisition of Agri-Biotech Applications (ISAAA). Brief 38 (www.isaaa.org).

CASE STUDY 4

VIRUS-RESISTANT PAPAYA

PREAMBLE

The introduction of viral sequences into a host plant genome has been shown to confer resistance to the cognate virus. The initial experiments showed that the expression of the viral coat protein gave resistance; this is termed coat protein protection or pathogen-derived resistance.

This case study examines two specific issues:

- The transgenic plants were made by a public organization, Cornell University, USA and not, as most other released transgenic crops, by a private company; some parts of the constructs were obtained from a private organization.
- Viruses are considered to be pests. The US Environment Protection Agency regulates the use of pesticides whether they are applied externally or produced within the plant (plant-incorporated protectants, termed PIPs). This raises questions as to whether a viral coat protein (which is not infectious) should be considered to be a PIP.

I. BACKGROUND

Some of the major points in the proposal are outlined below; more details can be found in the references at the end of this case study.

A. The Problem

Papaya production is an important industry for resource-poor farmers in Hawaii. In 1992 there was an outbreak of *Papaya ringspot virus* (PRSV) infection in the Puna district of Big Island, Hawaii where about 95% of the Hawaiian production acreage is located. PRSV causes a severe disease leading to death of the plant. It is spread rapidly by aphids and the only control measures are the application of insecticides and by roguing infected plants. These only delay the spread of the disease and, once introduced, PRSV has never been successfully eradicated from any region. There is no true resistance to PRSV in papaya germplasm and only some tolerance, which is moderately effective and is polygenic in inheritance.

PRSV belongs to the *Potyvirus* genus and has several distinct strains in different parts of the world. The genetic information of its RNA genome is expressed as a single protein (a polyprotein) which is cleaved into the viral gene products by virus-encoded proteases (enzymes which cleave proteins at specific sites). The RNA genomes of several strains have been sequenced and the viral proteins characterized.

II. PROPOSAL

The application was for the release of two lines of papaya cultivar Sunset (55-1 and 63-1) transformed with a sequence expressing the coat protein of PRSV. The two lines had slight differences in the inserted DNA. The information supplied to the regulatory authorities for risk assessment is summarized below.

A. The Recipient

Papaya (*Carica papaya*) of the genus *Carica* is a perennial tree widely grown in tropical regions for the production of fruit and of the proteolytic enzyme, papain (used for tenderizing meat). It is a polygamous species having a mating system that is either dioecious (staminate and pistliiate plants) or gynodieocious (hermaphrodite and pistillate plants). In commerce gynodieocious lines are preferred as they have the potential for inbreeding and consequent uniformity. Pollination of papaya flowers is considered to be relatively unspecialized and occurs through both insect and wind dispersal. Hermaphrodite plants have the capacity for self-pollination which is enhanced by anthesis occurring slightly before the flower opens. *C. papaya* does not naturally intercross with other *Carica* species, but some hybrids have been produced using embryo rescue techniques.

Wild papayas are uniformly dioecious and have weedy characteristics such as prolific seed production, minimal edible flesh and seed dormancy. They are found from southern Mexico to northern Honduras and throughout the Caribbean islands, but not in Hawaii. The cultivated papaya in Hawaii has never been listed by the Department of Agriculture as a noxious weed.

The recipient for transformation is *C. papaya*, cultivar "Sunset" which is a gynodieocious line of commercial importance in Hawaii.

B. The Donor

The coat protein gene was derived from PRSV mild mutant strain HA 5-1 produced by nitrous acid treatment of the severe Hawaiian strain PRSV HA. The 3'-terminal region of the genomic RNA of PRSV HA 5-1, including the coat protein gene has been sequenced.

C. The Constructs

The coat protein gene of PRSV HA 5-1 was isolated and ligated into the transformation vector pGA482GG (derived from the *Agrobacterium* binary vector, pGA482) (see Case Study 5 for binary vectors) to give the plasmid pGA482GG/PRV-4 (Fig. 1).

The PRSV coat protein construct in pGA482GG/PRV-4 is between the left and right borders of the *Agrobacterium* T-DNA, and is expressed as a chimeric protein with codons specifying the first 16 amino acids of *Cucumber mosaic virus* (CMV) coat protein which were fused to the PRSV HA 5-1 coat protein. This is because PRSV coat protein gene is part of a polyprotein and, thus, its mRNA does not have 5' transcriptional and translational elements necessary for expression; these are supplied by the CMV sequences. The genetic components of pGA482GG/PRV-4 are listed in Table 1.

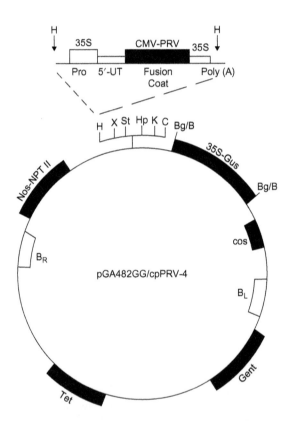

FIGURE 1 *Agrobacterium* binary vector pGA482GG/PRV-4 which was constructed by cloning a HindIII fragment containing the chimeric CMV/PRSV gene with promoter and terminator into pGA482. The location of important genetic elements are indicated: B$_R$, right border; B$_L$ left border; Nos-NPTII, plant expressible neomycin phosphotransferase gene; 35S-Gus, plant expressible GUS gene; cos, cos site; Tet, bacterial tetracycline resistance gene; Gent, bacterial gentamycin resistance gene. From Ling et al. (1991).

TABLE 1 Components of plasmid pGA482GG/cpPRV-4 (see Fig. 1) used in the development of the transgenic papaya lines 55-1 and 63-1

Item	Brief description
nos	Nopaline synthase promoter originally from *Agrobacterium tumefaciens*
nptII	Neomycin phosphotransferase originally from *Eschericia coli*
35S Pro	Promoter from *Cauliflower mosaic virus* (CaMV)
5'-UT	70 bp 5' untranslated region of *Cucumber mosaic virus* RNA 3
CMV-PRV fusion coat	Coat protein of PRSV HA 5-1 which has codons specifying the first 16 amino acids of CMV coat protein at its N-terminus
35S Poly (A)	Poly (A) terminator from CaMV
GUS	B-glucuronidase gene originally from *E. coli*
Gent	Gentamycin resistance gene (referred to as GENT^R) originally from *E. coli*
Tet	Tetracycline gene (referred to as TET^R) originally from *E. coli*

D. Transformation Method

The plasmid pGA482GG/PRV-4 coated onto tungten particles was transformed into embryogenic cultures of papaya cultivar "Sunset" by the biolistic microprojectile process (see Chapter 1, Section III.C.1).

E. Analyses of Products

1. Southern Blot Analyses

Southern blots analyses were performed on lines 55-1 and 63-1 using probes prepared from restriction enzyme products within the T-DNA borders. These probes were for nptII, GUS and PRSV coat protein. In addition, Southern blot analyses were conducted using probes to the gentamycin resistance gene, to the Ori V/Tet region together with about one-third of the tetracycline resistance gene, and to the Ori T/ Tet region together with about two-thirds of the tetracycline resistance gene, to determine if any fragment outside the T-DNA borders was incorporated into the genome of the transgenic lines during bombardment. Table 2 summarizes the results of these analyses.

Table 2 shows that the parts of the construct DNA that were inserted differ between lines 55-1 and 63-1. Line 55-1 did not have a gentamycin resistance gene or a complete tetracycline resistance gene. Line 63-1 did not contain the GUS gene (supported by lack of positive GUS assays in plant tissues). Although the probes hybridized with the gentamycin and tetracycline resistance genes it is unlikely that these would be expressed in plants as they have prokaryotic promoters.

2. Expression and Mendelian Inheritance of Transgenes

Transformed plants were assayed for GUS expression using histological and fluorimentric assays and, for nptII and coat protein expression, using enzyme-linked

TABLE 2   Summary of Southern blot analysis results

Probe	Control	55-1	63-1
Inside T-DNA borders			
Coat protein	−	+	+
GUS	−	+	−
nptII	−	+	+
Outside T-DNA borders			
Gentamycin	−	−	+
Ori T/Tet	−	+	+
Ori V/Tet	−	−	+

immunosorbent assays. R0 plants of line 55-1 expressed all three transgenes and R0 plants of 63-1 expressed only the nptII and coat protein genes; these results confirmed the Southern blot analyses (Table 2).

Inheritance of transgenes was examined in the R1 generation by analysing segregation ratios in 1- to 2-month-old seedlings from crosses of R0 transgenic plants. Crosses of R0 55-1 with non-transgenic "Sunset" produced progeny that conformed well for all three transgenes, with a ratio of 1 transgenic:1 non-transgenic plant indicating a single transgene insertion site (Table 3).

A transgenic 46-1 R0 plant, which expressed only the *nptII* gene, was crossed as female parent to a 63-1 R0 plant, to produce 46-1 × 63-1 progenies. The observed segregation ratio for coat protein was a poor fit to the expected 1 coat protein +:1 coat protein-ratio, but the nptII segregation was very close to the expected 3 npt II +:1 npt II-ratio (Table 3).

Expression of transgenes was measured in R2 and R3 generations of 55-1. The segregation of the coat protein gene in R2 seedlings from self-pollinated 55-1 R1 plants followed the 3 CP + :1 CP-ratio expected from a single transgene insertion site, but GUS expression was less frequent than expected. However, GUS expression was normal in the R3 generation (Table 3).

3. Resistance to PRSV
Both lines 55-1 and 63-1 were resistant to the Hawaiian strain of PRSV.

4. Disease and Pest Characteristics of Transgenic Papaya
A potential risk is that a viral transgene or its product could interact with a superinfecting virus which the transgenic plant is not protected against, and could cause a new disease or even a new virus. Three scenarios have been suggested (Hull, 1990): encapsidation of the genome of the superinfecting virus by the transgenically-expressed coat protein; synergistic interaction between the transgenically-expressed protein and the superinfecting virus to give a more severe disease; recombination between the transgene and the genome of the superinfecting virus resulting in a new virus. All three scenarios were considered and

TABLE 3   Inheritance of transgenes in papaya lines 55-1 and 63-1 (see text for crosses)

Line	n	Expected ratio	GUS	nptII	CP	X^2
R1 generation						
55-1 × "Sunset"	394	197:197	193:201	193:201	193:201	0.16
63-1 × 46-1	60	30:30			39:21	5.40
63-1 × 46-1	60	30:30		46:14		0.09
Leaves of R2 plants						
55-1 (+/CP) selfed	323	242:81			237:86	0.45
55-1 (+/CP) selfed	279	209:70	157:122			52.19
R3 seed (embryo and endosperm)						
55-1 (GUS/GUS) selfed	1410	1410:0	1410:0			0.0
55-1 (+/gGUS × (+/+)	1290	645:645	625:665			1.24
55-1 (+/GUS) selfed	285	214:71	224:61			1.88

n = number tested; X^2 = chi squared

thought to be no more likely than that resulting from joint infection of PRSV and the superinfecting virus. Thus, it was concluded that the transgenic papaya had no plant pest characteristics.

5. Nutritional Composition

The transgenic papaya developers analysed vitamins A and C and total soluble solids (a measure of the sugar content), which are considered to be significant nutrient components of papaya. They showed that the levels of these compounds were within the range obtained for several commercial varieties of papaya.

6. Toxicants

The presence of benzyl isothiocyanate (BITC) which is associated with latex in non-transgenic green papaya has been linked to prostate cancer in men over 70 years old and some cases of abortion in pregnant women. No significant differences were found in the BITC content of immature (green) and ripe transgenic fruit and non-transgenic fruit; the ripe fruit normally lacks latex and has virtually no BITC. It was concluded that ripe transgenic papaya fruits posed no special threat to human health.

E. Environmental Consequences of the Introduction of Transgenic Papaya

1. Weediness

Since papaya does not intercross naturally with other *Carica* species, and since no other *Carica* species is considered to be a weed, there was no evidence to suggest that PRSV resistance resulting from gene flow from transgenic papaya would cause weediness problems.

2. Native Floral and Faunal Communities

No direct pathogenic properties, or any hypothetical mechanisms for pathogenesis towards beneficial organisms such as bees and earthworms, were identified for either of the transgenic papaya lines. No other potential impacts with agricultural or natural ecosystems could be identified.

III. RISK ASSESSMENT

The risk assessment was undertaken by the United States Department of Agriculture, Animal and Plant Health Inspection Service (USDA–APHIS) for environmental aspects and by the US Food and Drug Administration (FDA) and the Environmental Protection Agency (EPA) for food aspects (see Chapter 6, Table 6.2). Also APHIS solicited written comments on the proposal from the public. During the designated 60-day comments period, 18 comments were received from universities, papaya growers and processors, a papaya industry association, an office of cooperative extension service and a state department of agriculture; all comments were favourable to the application.

IV. DETERMINATION

A. Environmental Aspects

Based on analysis of the data submitted by Cornell/Hawaii, a review of other scientific data, comments received and field tests, APHIS has determined that papaya lines 55-1 and 63-1:

- Exhibit no plant pathogenic properties;
- Will not increase the likelihood of the emergence of new plant viruses;
- Are no more likely to become weeds than papaya developed by traditional breeding techniques;
- Will not increase the weediness potential for any other cultivated or wild species with which they can interbreed;
- Will not harm threatened or endangered species or other organisms, such as bees, that are beneficial to agriculture;
- Will not cause damage to processed agricultural commodities.

Therefore, APHIS has concluded that the subject papaya lines and any progeny derived from hybrid crosses with other non-transformed papaya varieties will be as safe to grow as papaya in traditional breeding programmes.

B. Food Aspects

The developers of the transgenic papaya have concluded, in essence, that the PRSV-resistant papaya they have developed is not materially different, in terms of food safety and nutritional profile, from red-pigmented papaya varieties with a history of safe use. Although the two transgenic papaya lines contain sequences of antibiotic resistance genes, some of which are still in clinical practice, (Table 2) these were not considered to be a significant problem. At this time, based on the papaya developers' description of their data and analyses, the FDA and EPA consider the consultation on the virus-resistant transgenic papaya line 55-1 to be complete.

V. GENERAL ISSUES

A. Virus Protein

As noted above, the expression of viral genes to induce protection against the cognate virus has the potential of leading to either the enhancement of a viral disease or the creation of a new virus. In the case of PRSV coat protein expression in papaya it was concluded that this potential risk was not significant. The comparator of any possible interactions between a viral transgene and a superinfecting virus is the natural infection of that host by donor virus of the transgene and the superinfecting virus. This should be taken into account in risk assessing the release of a crop plant expressing a virus transgene.

B. Durability

There are several strains of PRSV which differ in genome nucleic acid sequence. After the release in Hawaii, it was found that the transgenic lines were not protected against the South East Asia strains of the virus – new transformants are being made to obtain coat-protein protection against these strains. Thus, it is important to have an understanding of the range of variation of a target virus in the deployment of virus-resistant crops containing virus transgenes.

C. Release of Transgenic Papaya in Other Countries

As noted in Section II.A, wild papayas are found in Central America, and throughout the Caribbean islands. If it was proposed to release transgenic PRSV-resistant papayas in these regions consideration should be given to the possibility of gene flow from the GM plants to the wild plants. One of the factors that would need to be ascertained would be whether PRSV infection controlled the vigour of the wild plants, and whether the introduction of resistance into these plants would affect their status in the ecosystem.

D. Plant-incorporated Protectants (PIPs)

As noted above, the EPA regulates the use of PIPs, a responsibility given to them some time before the use of viral sequences in transgenic plants. They were faced with having to make a determination as to whether a viral transgene was a PIP. This conundrum illustrates some of the problems, such as issues that might have questionable specific relevance to the GM regulations, which might arise from the adaptation of previous legislation to the regulation of GMO releases.

E. Intellectual Property

Although not a risk assessment issue, a major constraint on the release of the transgenic papaya was the intellectual property (IP) rights of some of the constructs and techniques used by the public organization that used the lines. The IP rights were owned by major biotechnology companies and detailed negotiations had to take place to reach a solution so that the transgenic papaya could be used by resource-poor farmers (Mendoza and Botella, 2008), http://www.apsnet.org/education/feature/papaya

References and Further Information

Gonsalves, D. (1998). Control of papaya ringspot virus in papaya: A case study. *Annu. Rev. Phytopathol.* **36**, 415–437.

Hull, R. (1990). Non-conventional resistance to viruses in plants: concepts and risks. In *Proceedings of the 19th Stadler Conference*, (G.P. Gustafson, ed.). Plenum Press, New York, pp. 443–457.

Ling, K., Namba, S., Gonsalves, C., Slightom, J.L. and Gonsalves, D. (1991). Protection against detrimental effects of potyvirus infection in transgenic tobacco plants expressing the papaya ringspot virus coat protein. *Bio/Technology* **9**, 752–758.

Mendoza, E.M. and Botella, J.R. (2008). Recent advances in the development of transgenic papaya technology. *Biotechnol. Annu. Rev.* **14**, 423–462.

http://www.agbios.com/dbase.php?action = ShowProd&data = 55-1%2F63-1 and references therein.

http://64.26.172.90/docroot/decdocs/9605101.htm.

CASE STUDY 5

REDUCTION OF RIPENING GENE EXPRESSION OF TOMATOES: NO PROTEIN PRODUCED BY TRANSGENE

PREAMBLE

There have been two releases of tomatoes (Flavr Savr produced by Calgene and Long Life Tomato TGT7-F produced by Zeneca) which have been modified to reduce the expression of a plant gene that controls ripening. The background to these lines is described in Chapter 1, Box 1.9. The transgene in both these lines targeted the plant enzyme, polygalacturonidase (PG), a pectinase which hydrolyses the pectin chains in the cell wall. This hydrolysis is an important part of the ripening process and in changing the viscosity characteristics of processed tomato.

In Flavr Savr, a construct expressing an antisense RNA to the messenger RNA (mRNA) of the enzyme polygalacturonidase (PG) was inserted into tomato. The antisense RNA does not express any protein, but binds to the PG mRNA restricting its expression and slowing the ripening of the tomato. The actual mechanism by which the transgene interfered with the expression of the PG gene is not fully understood. One possibility is that it hybridized to the PG mRNA, preventing its translation. Another possibility is that, by forming a double-stranded RNA, it induced RNA silencing defence mechanism of the plant (see Chapter 1, Box 1.6 for RNA silencing) or it might be a combination of the two mechanisms.

A sense copy of a shortened PG gene was inserted into line TGT7-F. This insert did not express a protein and its presence reduced the activity of the endogenous PG. At the time of producing, testing and marketing the line, the mechanism by which the expression of PG was affected was not known. Subsequently, it was found to operate through an RNA silencing mechanism (Han and Grierson, 2002).

The important feature of this case study is that the transgenes did not express any protein. This situation is different to those in the other case studies, and raises the question as to whether the risk assessment considerations should be different. In the future it is likely that there will be many more transgenic lines using the RNA silencing strategy, either to give resistance to pests and pathogens, or to achieve a new trait by tailoring the plant metabolic pathways.

It should be noted that the determinations were made on these releases in 1992–1994 for Flavr Savr as whole fruit in the US, and in 1998 for TGT7-F as canned processed fruit in the UK. Detailed regulations have changed since that time, especially in Europe, but factors taken into account in the appraisals were very similar to those that would be used now.

I. THE PROPOSALS

Some of the major points in the proposal are outlined below; more details can be found in the references at the end of this case study.

A. Flavr Savr

1. The Recipient

The recipient was tomato cultivar MacGregor. The regulatory authorities were supplied with detailed information on the biology and other properties of unmodified tomato (see www.agbios.com/docroot/decdocs/05-242-001.pdf).

2. The Construct

The construct was assembled into disarmed Ti plasmid which did not contain genes for phytohormone or tumor formation. A map of the inserted DNA is shown in Fig. 1, and details of its genetic components are given in Table 1.

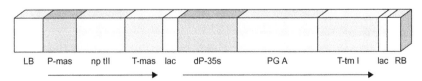

| LB | P-mas | np tII | T-mas | lac | dP-35s | PG A | T-tm I | lac | RB |

FIGURE 1 Details of the construct used to make Flavr Savr tomato. Abbreviations are defined in Table 1.

TABLE 1   Genetic components of antisense PG inserted construct

Abbreviation	Element name	Source	Size (KB)
LB	Left border	*Agrobacterium tumefaciens* T-DNA	0.58
P-mas	Mannopine synthase promoter	*A. tumefaciens*	0.68
nptII	Neomycin phosphotransferase	*Escherichia coli* transposons Tn5	0.98
T-mas	Mannopine synthase terminator	*A. tumefaciens*	0.77
lac	Beta-galactosidase	*E. coli*	0.29
dP-35S	Double 35S promoter	*Cauliflower mosaic virus*	1.2
PG A	Antisense polygalacturonidase gene	Tomato	1.6
T-tml	tml gene and transcript 7 gene	*A. tumefaciens* Ti plasmid pTiA6	1.2
lac	Beta-galactosidase	*E. coli*	0.17
RB	Right border	*A. tumefaciens* T-DNA	0.28

4. Transformation Method

Transformation was by the *Agrobacterium*-mediated method (see Chapter 1, Box 1.7) using a binary vector (see Bevan, 1984), the *vir* genes (necessary for the transfer of the T-DNA) being on a separate non-integrating plasmid to that containing the construct.

5. Analyses of Products

a. Southern Blot Analyses   Southern blot analysis showed that a single copy of the complete T-DNA region encoding the antisense sequence to the PG gene and the *nptII* gene was inserted, and that there was no incorporation of plasmid DNA sequences outside the T-DNA region.

b. Expression and Mendelian Inheritance of Transgenes   The sequences in the T-DNA were inherited in a dominant Mendelian manner and were stable for at least five generations. The only novel protein expressed was nptII. The antisense PG sequence did not express any protein and its presence reduced the endogenous PG activity to less than 1% of the activity found in the unmodified parental line.

c. Biology and Disease and Pest Characteristics of Flavr Savr Tomato   The biology of the Flavr Savr tomato was essentially the same as that of the parent MacGregor tomato. Similarly, the transgenic line did not differ from the unmodified parent in disease and pest characteristics.

6. Environmental Factors

Extensive field testing of Flavr Savr tomato conducted over five years demonstrated that the variation in its agronomic characteristics was within the range of variation of unmodified tomatoes. The tests showed no differences in outcrossing, weediness characteristics, in effects on non-target organisms, biodiversity or on the general environment.

7. Food and Feed Considerations

Food considerations of Flavr Savr tomato focused on the fact that tomatoes can form a significant part of the diet, and that they are often eaten in an uncooked form. The information supplied to the regulators is summarized in www.agbios. com/docroot/decdocs/bnfMFLV.pdf.

No significant differences were found between Flavr Savr and its unmodified comparator in nutritional composition, or in its toxicity or allergenicity.

8. Risk Assessment and Determinations

Based on the information supplied by the applicant to the US regulatory authorities (Animal and Plant Health Inspection Service of the Department of Agriculture and the Food and Drug Administration of the Department of Health and Human Services) authorization was given for the environmental release of Flavr Savr tomato and its use for food and feed.

B. Zeneca TGT7-F Event

Information on this line can be found at http://www.ncbe.reading.ac.uk/NCBE/GMFOOD/tomato.html

1. The Recipient

The recipient was tomato cultivar TGT7.

2. The Donors

The PG sequence was the 731 bp fragment (approximately 50% of the gene) from the 5' end of a PG cDNA clone from the Ailsa Craig variety of tomato (Smith et al., 1990).

The *nptII* selection marker gene was from the Tn5 transposon of *E. coli*, strain K12.

3. The Construct

The construct was by a sense orientation of the truncated PG gene between the *Cauliflower mosaic virus* 35S promoter and the 3' end of the nopaline synthesase gene (giving the terminator) and cloned into the binary vector Bin 19 (Bevan, 1984) which contains the *nptII* selection marker gene.

4. Transformation Method

Transformation of stem internode explants excised from tomato seedlings (Ailsa Craig) was by the *Agrobacterium*-mediated method (Bird *et al.*, 1988).

5. Analyses of Products

a. Southern Blot Analyses   Southern blots showed that a single complete copy of the T-DNA containing the truncated PG sequence and the *nptII* gene was inserted into the recipient.

b. Expression and Mendelian Inheritance of Transgenes   No protein was detected to be expressed in tomato line TGT7-F from the truncated PG sequence even in leaves – endogenous PG is only found in ripening fruits. The PG activity in ripening fruits was reduced to about 1% of that in untransformed controls. nptII protein was the only new protein detected in line TGT7-F.

c. Biology and Disease and Pest Characteristics   The disease, pest, and other agronomic characteristics of the transgenic line did not differ from those of non-transgenic lines.

6. Environmental Factors

Field testing of TGT7-F tomato showed that the variation in its agronomic characteristics was within the range of variation of unmodified tomatoes. The tests showed no differences in outcrossing, weediness characteristics or effects on non-target organisms, on biodiversity or on the general environment.

7. Food and Feed Considerations

a. Compositional Analysis   The only difference between line TGT7-F and the parental variety was the size distribution of pectin molecules in the fruit and in the small amount of nptII protein. The pectin chain length in the processed products of fruits of TGT7-F was well within the range commonly found in processed products of traditional tomatoes.

b. Nutritional Composition   There were no statistical differences between fruits from TGT7-F and its comparator in contents of vitamins A and C, calories, fat, sodium, carbohydrate, fructose, glucose, dietary fibre, protein, calcium and iron. There were also no statistical differences in insoluble and soluble fibre content, ash, titratable acidity, pH or colour (measured on the Hunter color L:A:B scale).

c. Toxicity and Allergenicity   The glycoalkaloids, chaconine and solanine, could not be detected in line TGT7-F; the levels of the glycoalkaloid tomatine were equivalent to those found in comparator non-GM tomatoes. Analysis of the levels of biogenic amines (tyramine, tryptamine and serotonin), which are considered as potential toxins in tomato, did not reveal any significant differences between the transgenic line and non-transgenic parental line.

8. Risk Assessment and Determinations

The UK regulators concluded that the transgenic line was substantially equivalent to non-transgenic tomatoes, apart from reduced PG activity and the expression of nptII. They did not consider that either of these differences posed any risk to food and feed or to the environment. Similar conclusions were reached by the US and Canadian regulatory authorities.

C. Considerations on These Two Releases

1. Although these risk assessments were made 10–15 years ago, the data supplied to the regulatory authorities was very similar to that which would be used for a current application.

2. The only major difference if a risk assessment was to be made on these products now would be a further consideration of the presence of an antibiotic resistance marker gene. This would not be permitted in several regulatory authorities, especially the EU. However, as noted in Chapter 1, Box 1.9, the production of Flavr Savr tomato was discontinued because of commercial considerations. The Zeneca canned tomatoes were initially a great success, but were withdrawn due public concern about GMO products in general.

3. Labelling of Zeneca canned tomatoes (Fig. 2) stated that product was made with genetically modified tomatoes, although there were no legal requirements to do so. This was the first time a GM food had been released in Europe, and it sold very rapidly at a lower price that non-GM canned tomato.

FIGURE 2   Can of GM tomato puree (TGT7-F) showing label indicating that it was made with "genetically modified tomatoes." (see colour section).

II. OTHER SITUATIONS AND ISSUES

A. Regulation of Transgenes that do not Express Proteins

As there have not yet been extensive releases of antisense/RNA silencing transgenic crops there is not the familiarity with this strategy. As more such events reach the risk assessment stage, regulatory authorities will have to develop approaches that recognize that some of the potential risks of transgenes expressing proteins would not be relevant. Among the major considerations for non-protein genes would be:

- Environment:
  - The trait is unlikely to affect non-target organisms, but could affect the natural ecology due to gene flow. An example of this would be disease resistance affecting the dominance of sexually-compatible species in natural ecosystems, which is similar to the situation with a protein-based disease resistance trait. A baseline for this would be conventionally-bred disease resistance;
  - The situation with gene stacking would be similar to that of protein-expressing transgenes (see Case Study 2);
  - The process of transformation or the expression of the non-protein transgene might affect the plant metabolism creating a new phenotype.
- Food and feed:
  - As some transgenes may be directed at changing the plant metabolism to create a new trait (e.g., drought resistance, altered starch or fatty acid composition) there may be unintended effects on linked metabolic pathways, leading to changes in nutritional characteristics or the production of a new toxin or allergen. The potential for such risks will be easier to assess with more familiarity with such metabolic changes.

The suggestion is made in Chapter 1, Section III.C that inserts that do not express a protein and function by RNA silencing or antisense should be termed "silencing genes," to contrast them from inserts that produce proteins which should be termed "conventional genes."

References and Further Information

Bevan, M. (1984). Binary *Agrobacterium* vectors for plant transformation. *Nucleic Acids Res.* **12**, 8711–8721.

Bird, C.R., Smith, C.J.S., Ray, J.A., Moureau, P., Bevan, M.W., Bird, A.S., Hughes, S., Morris, P.C., Grierson, D. and Schuch, W. (1988). The tomato polygalacturonidase gene and ripening-specific expression in transgenic tomato. *Plant Mol. Biol.* **11**, 651–662.

Han, Y. and Grierson, D. (2002). Relationship between small antisense RNAs and aberrant RNAs associated with sense transgene mediated gene silencing in tomato. *Plant J.* **29**, 509–519.

Smith, C.J.S., Watson, C.F., Bird, C.R., Ray, J., Schuch, W. and Grierson, D. (1990). Expression of a truncated tomato polygalacturonidase gene inhibits expression of the endogenous gene in transgenic plants. *Mol. Gen. Genet.* **224**, 477–481.

http://www.agbios.com/dbase.php?action = ShowProd&data = FLAVR + SAVR and references therein.

Index

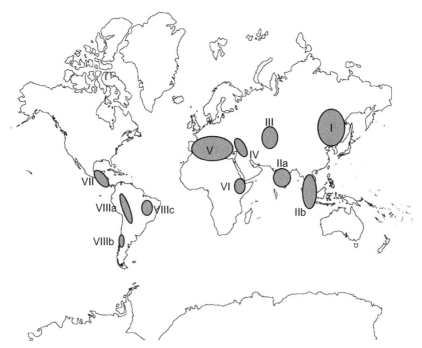

PLATE 1   Vavilov's primary centres of origin of the world's major crops.

PLATE 2 Variation in *Brassica oleracea*. In the centre is the wild-type *B. oleraces* surrounded by cultivated variants with enlarged terminal bud (cabbage), coloured variety of swollen terminal bud, swollen axillary buds (Brussels sprouts), swollen flower heads (cauliflower and broccoli) and swollen stem (kohl rabi).

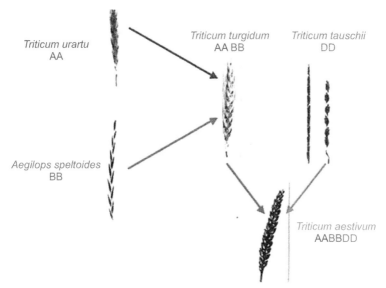

PLATE 3 Evolution of modern wheat (Triticum aestivum) arising from natural inter-specific crosses of wild grass species bringing together three genomes.

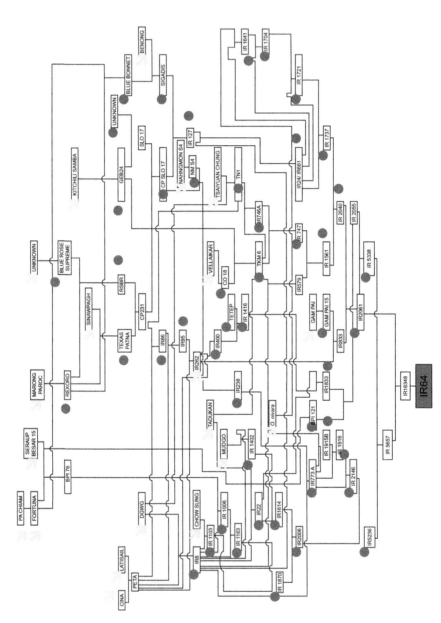

PLATE 4 Breeding tree for rice variety IR64. The boxes show the rice species and varieties used in the breeding programmes. The yellow arrows indicate landraces and the red dots indicate new varieties derived either from breeding programmes or by mutagenesis.

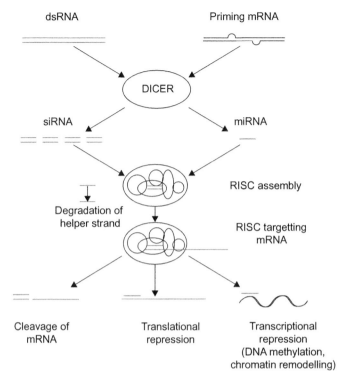

PLATE 5 RNA silencing pathways. The red line indicates the guide strand and the green line the anti-guide (helper) strand and the target mRNA. The left-hand pathway is that for mRNAs and the right-hand pathway for micro-RNAs which are involved in plant development. Both pathways pass through Dicer which cleaves the double-stranded RNA to small fragments and through RISC which amplifies the system. From Hull (2009) with kind permission of Elsevier.

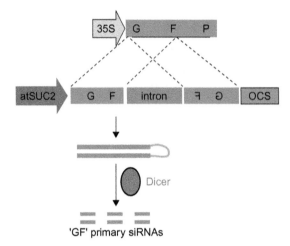

PLATE 6 Construct for RNA silencing gene. This example shows a construct for silencing the expression of green fluorescence protein (GFP). The top line shows the GFP gene used as a marker in a plant. The second line shows the silencing construct with a promoter (red arrow on left), portions of the GFP sequence in positive and reverse sense separated by an intron and on the right the terminator. The next two lines show how the RNA expressed from the construct forms a double-stranded molecule which is processed to form siRNAs by Dicer.

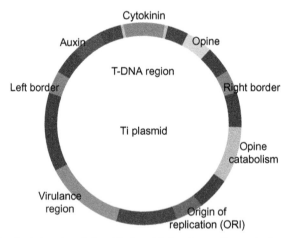

PLATE 7 *Agrobacterium tumefaciens*. Ti plasmid showing coding and other regions. At the top the T-DNA region, bounded by the left and right borders, is indicated. In the unmodified plasmid this region contains genes for the maintenance of *A. tumefaciens* in its host; these genes are removed to make the plant transformation vector. The lower part of the genome contains genes and regions involved in the replication of the plasmid and in the process of plant infection by the bacterium. From http://upload.wikimedia.org/wikipedia/commons/8/89/Ti_Plasmid.jpg.

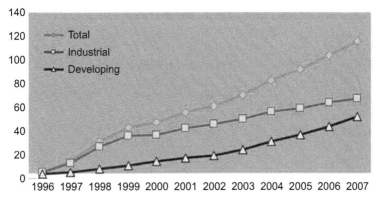

A

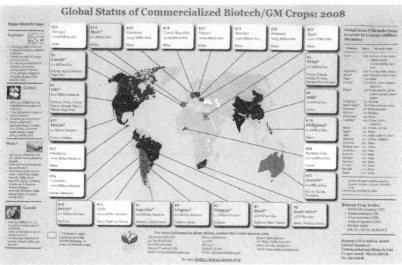

B

PLATE 8 **A**. Global area of biotech crops 1996–2008: industrial and developing countries (million hectares). **B**. Countries growing biotech crops in 2007, 13 of which are termed "mega-countries", growing 50,000 hectares or more. From James (2007; 2008). With kind permission of the International Service for the Acquisition of Agrobiotech Applications.

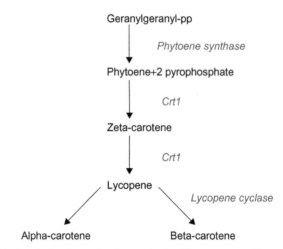

PLATE 9 Biochemical pathway for the synthesis of beta-carotene. The enzymes involved are shown in red.

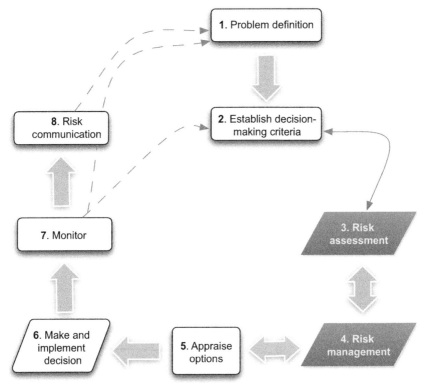

PLATE 10   **1. Scope of the assessment** – Field trial; Food and feed safety; Environmental release; Commercial release; Import. **2. Establish decision-making criteria** – What are the regulatory requirements? Set up rules for decision making. **3. Risk assessment** – Identify risk source(s) and susceptible biota; Exposure scenarios; Assessment endpoints; Estimate magnitude of risk. **4. Define the risk management plan** – Consider alternative risk management options; Consider monitoring requirements. **5. Appraise risk management and monitoring options** – Evaluate options in terms of feasibility and cost effectiveness. **6. Decision making** – Were the decision-making criteria adequate? Has the problem been defined correctly? **7. Implementation of the monitoring plan** – Does the information gathered require amendment of the plan or redefinition of the problem? **8. Communicate risk** – Establish criteria for stakeholder participation; Actively engage different stakeholder groups; Do stakeholder inputs require redefinition of the problem?

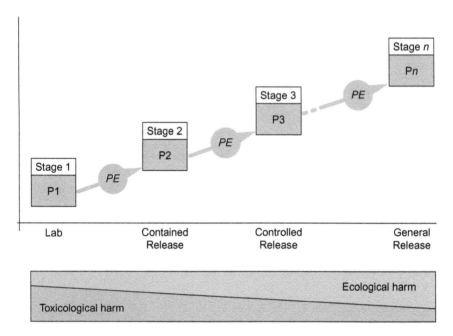

PLATE 11 Schematic representation the stages of ecological risk assessment for the different containment situations (laboratory to general release) starting with the formulation of the initial problem (P1) followed by a succession of Project Evaluations (PE) until the general release of the GM crop (P*n*). Each PE includes: identification of additional problems; modelling; problem analysis; conclusion; and decision. Concomitant to the progression of the containment situations, the level of ecological harm rises whereas the level of toxicological harm falls. The figure is modified from an original that was prepared by Dr Jeffrey D. Walt for the e-Biosafety International Training Network (http://binas.unido.org/moodle).

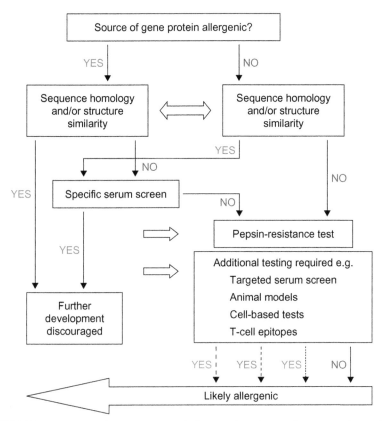

PLATE 12 Flow chart for assessment of allergenicity of a newly expressed protein in a GM organism. Note that the red lettering of "Yes" indicates potential allergenicity and fainter shades of red indicate less potential (from Davies, 2005).

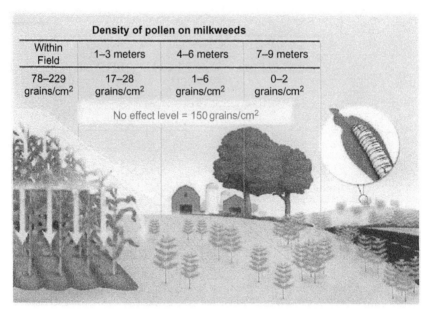

Density of pollen on milkweeds			
Within Field	1–3 meters	4–6 meters	7–9 meters
78–229 grains/cm^2	17–28 grains/cm^2	1–6 grains/cm^2	0–2 grains/cm^2
No effect level = 150 grains/cm^2			

PLATE 13 Density of pollen on milkweeds. (Preliminary data reported at the Monarch Butterfly Research Symposium by Dr. Mark Sears, University of Guelph, Dr. Galen Dively, University of Maryland and Dr. Rich Hellmich, USDA).

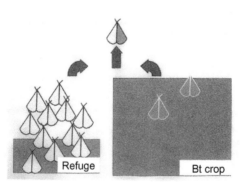

PLATE 14 Refugia strategy. Resistant insects (red wings) in the Bt crop mate with the susceptible insects (yellow wings) in the refuge to give heterozygous progeny. The progeny from any insects from the Bt crop with a recessive resistance gene (the usual situation) will be susceptible.

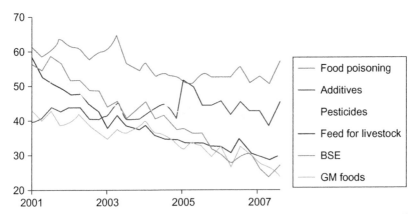

PLATE 15 Public concern with various issues concerning food. From UK Cabinet Office (2008). With kind permission of the UK Cabinet Office.

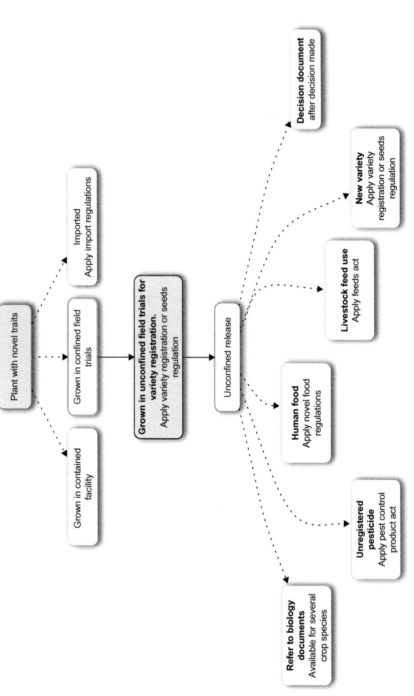

PLATE 16 Diagram showing the interlinking of considerations in the Canadian Novel Foods Regulations.

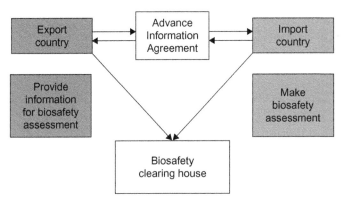

PLATE 17 Structure for the transference of biosafety information.

PLATE 18 Can of GM tomato puree (TGT7-F) showing label indicating that it was made with "genetically modified tomatoes.

Printed and bound by CPI Group (UK) Ltd, Croydon, CR0 4YY

03/10/2024

01040414-0004